Calculating and Problem Solving Through Culinary Experimentation

While many books proliferate elucidating the science behind the transformations during cooking, none teach the concepts of chemistry and physics through problem solving based on culinary experiments as does this one by a renowned physical chemist and one of the founders of molecular and physical gastronomy. *Calculating and Problem Solving Through Culinary Experimentation* offers an appealing approach to teach experimental design and scientific calculations.

Given the fact that culinary phenomena need chemistry and physics to be interpreted, there are strong and legitimate reasons for introducing molecular and physical gastronomy in the scientific curriculum.

As in any scientific discipline, molecular and physical gastronomy is based not only on experiments (to observe the phenomena to be studied) and calculation (to fit the many data obtained by quantitative characterization of the studied phenomena), but also on theoretical work without which no real science is done, including refuting consequences of the introduced theories. Often, no difficult calculations are needed, and many physicists and chemists, in particular, take their first steps in understanding phenomena with very crude calculations. Indeed, they simply apply what they learned before moving to more difficult math.

In this book, students are invited first to make simple experiments in order to get a clear idea of the (culinary) phenomena that they will be invited to investigate, and then are asked simple questions about the phenomena, for which they have to transform their knowledge into skills, using a clear strategy that is explained throughout. Indeed, the book could have the title "problem solving based on experiments", and all this about food and cooking.

Key Features:

- Introduces readers to tips for experimental work
- Shows how simple scientific knowledge can be applied in understanding questions
- Provides a sound method ("strategy") for calculation in physics and chemistry
- Presents important definitions and laws for physical chemistry
- Gives confidence in one's calculation skill and problem-solving skills
- Explores chemical and physical phenomena that occur during cooking

A unique mix of culinary arts and correct calculations, this book is useful to students as well as professors in chemistry, physics, biology, molecular and physical gastronomy, and food science and technology (including food engineering).

Calculating and Problem Solving Through Culinary Experimentation

Hervé This vo Kientza

CRC Press
Taylor & Francis Group
Boca Raton London New York

CRC Press is an imprint of the
Taylor & Francis Group, an **informa** business

First edition published 2023
by CRC Press
6000 Broken Sound Parkway NW, Suite 300, Boca Raton, FL 33487-2742

and by CRC Press
4 Park Square, Milton Park, Abingdon, Oxon, OX14 4RN

CRC Press is an imprint of Taylor & Francis Group, LLC

Library of Congress Cataloging-in-Publication Data

Names: This, Hervé, author.
Title: Calculating and problem solving through culinary experimentation / Hervé This vo Kientza.
Description: First edition. | Boca Raton, FL : CRC Press, 2023. | Includes
bibliographical references and index.
Identifiers: LCCN 2022012382 (print) | LCCN 2022012383 (ebook) | ISBN
9781032287140 (hbk) | ISBN 9781032286501 (pbk) | ISBN 9781003298151 (ebk)
Subjects: LCSH: Food--Experiments. | Molecular gastronomy. | Chemistry--Mathematics.
Classification: LCC TX652.95 .T457 2023 (print) | LCC TX652.95 (ebook) | DDC
664/.07--dc23/eng/20220726
LC record available at https://lccn.loc.gov/2022012382
LC ebook record available at https://lccn.loc.gov/2022012383

ISBN: 978-1-032-28714-0 (hbk)
ISBN: 978-1-032-28650-1 (pbk)
ISBN: 978-1-003-29815-1 (ebk)

DOI: 10.1201/9781003298151

Typeset in LM Roman
by KnowledgeWorks Global Ltd.

Contents

Preface

"I am the dean of the students of France."

**—French chemist Michel Eugène Chevreul, pioneer of the
chemistry of fat, at the age of 100 years old**

I do not really understand the need for forewords, when books have an introduction. Of course, sometimes, a foreword is requested of a personality, who says why the book is interesting, but indeed I hate this idea, because this is an "authority argument": I prefer the readers (I always consider them as friends) make their own judgment.

How does one avoid this kind of foreword? One possibility, as here, is to describe the circumstances in which the book was created: for reasons that I will provide in Chapter 1, I want to thank here the many "younger friends" who worked in the Group of Molecular and Physical Gastronomy since 2000 and who tested all this material.

By "younger friends", I mean the students who trusted me enough to think they could usefully learn alongside me (internships, university programs, etc.). They are friends (friends, again friends: I want only friendly people around me) because they have the same interests as I do: the natural Sciences and their applications.

And so, this book is for "younger friends", for anyone who wants to learn... including me. Many of my younger friends told me that I am crazy to think that anyone can calculate as nightingales sing, and they are perhaps right; however, we need goals, in particular calculation goals, in order to move in a science and technology direction, don't we? Even if we will not reach the calculation skills of people such as Maxwell, Boltzmann, Gibbs, Dirac or Poincaré, we can improve, with proper training.

A special dedication is made to my friends Roisin Burke, Alan Kelly, and Christophe Lavelle, who reviewed a previous version of this text, during the preparation of the *Handbook of Molecular Gastronomy*. Working with them in order to produce this wonderful (*i.e.*, useful) book was a great pleasure because the load was so easily shared among all of us.

REFERENCES

Any sentence, any idea given in scientific text should be justified by (good) references to the scientific liitterature, and the above foreword makes no exception. But here, *stricto sensu*, there are no references (except for the *Handbook of Molecular Gastronomy, CRC Press, 2021*), and in particular because this is a foreword; but wait: why not? After all, should not we specifically avoid—because this is a foreword—giving justifications of what we say or write? Indeed, it is not a question of appearance, but rather giving references is always a good way of being sure of what we think ourselves.

More generally, the goal is to improve oneself, isn't it?

As well, this explanation about references allows me the opportunity to add that there will be different kinds of references, including those that:

- justify information that is given
- provide more general and useful sources of information
- indicate very good texts, articles or books, that I wish to recommend (let us share the wonder for beautiful works), and indeed if you look at the references, you will see comments, sometimes.

But enough with appetizers, let us banquet!

Bumstead HA. 1893. Josiah Willard Gibbs, *The American Journal of Science*, https://zenodo. org/record/1450126, last access 2022-01-11.

Burke R, Kelly A, Lavelle C, This vo Kientza H (eds). 2021. *Handbook of molecular gastronomy*, CRC Press, Boca Raton, USA.

Cercignani C. 2006. *Ludwig Boltzmann: the man who trusted atoms*, Oxford University Press, Oxford, UK.

Farmelo G. 2009. *The strangest man: the hidden life of Paul Dirac, quantum genius*, Faber and Faber, London, UK.

Mawhin J. 2004. *Henri Poincaré, a life in the service of science*, https://www.ams. org/notices/200509/comm-mawhin.pdf, last access 2022-01-11.

Tetley Glazebrook R. 1885. Maxwell, James Clerk, *Dictionary of national biography*, https://en. wikisource.org/wiki/Dictionary_of_National_Biography,_1885-1900/Maxwell,_James_Clerk, last access 2022-01-11.

About the Author

Hervé This vo Kientza is a physical chemist at Inrae and invited professor at AgroParisTech. He is the director of the AgroParisTech-Inrae International Centre for Molecular and Physical Gastronomy, the head of the Irae Group of Molecular Gastronomy, the scientific director of the Foundation Science & Culture Alimentaire, at the French Academy of Sciences, the president of the Educational Committee of the Institute for Advanced Studies in Gastronomy (University of Reims Champagne Ardennes/Le Cordon bleu), a member of the Steering Committee of the Académie d'Alsace, and a scientific advisor of the journal *Pour la Science* (the French edition of Scientific American).

Born the 5th of June 1955, Hervé This vo Kientza studied chemistry and physics at the École Supérieure de Physique et de Chimie de Paris (ESPCI Paris), as well as modern literature at University Paris IV. Then he worked for 20 years at Belin Publishing Company and at the scientific journal *Pour la Science* (the French Edition of *Scientific American*), where he was respectively editor and editor-in-chief. During this time, since the 24th of March 1980, he developed his work in molecular and physical gastronomy in his personal laboratory first, then at the Collège de France, where he had been given a laboratory by Jean-Marie Lehn (Nobel Prize for chemistry in 1987).

At the same time, Hervé This vo Kientza contributed many radio programs (*Panorama, France Culture*), and he was the scientific director of the science program Archimède (Arte) and Pi=3.14 (*France 5*). He also created and animated weekly TV programs on *France 5*, and daily radio programs for French radios *France Culture and France Inter*.

After 1980, while he was conducting his scientific research in what was later (in 1988) to be known as Molecular and Physical Gastronomy, Hervé This vo Kientra also promoted a new way of cooking that was called "Molecular Cooking" (that generated a cookery style called Molecular Cuisine), based on the use of modern tools from laboratories.

In 1988, Hervé This vo Kientza officially created the scientific discipline first called "Molecular and Physical Gastronomy", with Nicholas Kurti (FRS, 1908–1998, director of the Clarendon Laboratory, professor of physics at Oxford).

In 1995, upon the invitation of the President of the French Academy of Sciences, he defended a PhD in physical chemistry for which he was his own director, under the title *La gastronomie moléculaire et physique*. The jury included the two Nobel prize winners Pierre Gilles de Gennes and Jean-Marie Lehn, but also the famous chemist Pierre Potier. The next year, he was invited by Jean-Marie Lehn to have a laboratory at the College de France, and in 1999, Hervé This vo Kientza was invited to defend his habilitation to direct research, with a jury including Guy Ourisson (then President of the Académie des Sciences), Xavier Chapuisat (then president of University Paris Sud), Étienne Guyon (then director of the Ecole Normale Supérieure), Alain Fuchs (former president of CNRS) and the Three Star Chef Pierre Gagnaire.

In 2000, Hervé This vo Kientza left the companies *Belin* and *Pour la Science*, and devoted all his time on scientific research at INRA, first in his laboratory within the Department of Chemical Interactions of the Collège de France (director J-M Lehn).

Also in 2000, after the publication of his book *La Casserole des Enfants*, the French Minister of Public Education (Jack Lang) asked Hervé This vo Kientza to create and develop in all primary schools of France some specific programs called *Ateliers Expérimentaux du Goût*. These educational programs were first tested in the academies of Paris and la Réunion. They were officially launched in 2001. The same year, his book *Traité Élémentaire de Cuisine* was the basis for the new programs for French public culinary schools.

In 2004, these programs, and some others (*Dictons et Plats Patrimoniaux*, etc.) were followed by the *Ateliers Science & Cuisine* and are today used in high schools.

Also in 2004, the French Secrétaire d'État aux PME, au Commerce, à l'Artisanat, aux Professions libérales et à la Consommation asked Hervé This vo Kientza to create the Institut des Hautes Études du Goût, de la Gastronomie et des Arts de la Table (Advanced Studies in Gastronomy), with the Université de Reims Champagne Ardennes. He was appointed president of its educational program.

In April 2006, as he was moving his laboratory to AgroParisTech, the French Academy of Sciences invited Hervé This vo Kientza to create the Foundation Science & Culture Alimentaire, of which he was appointed the scientific Director.

Then, in 2010, Hervé This vo Kientza was asked to head the Food Section of the French Academy of Agriculture, which he did for nine years. And in June 2014, he was asked to create the AgroParisTech-Inrae International Centre for Molecular and Physical Gastronomy, of which he was appointed as a director.

Hervé this vo Kientza is also the director of the International Workshops on Molecular Gastronomy (since 1992), and Director of the Journées Françaises de Gastronomie Moléculaire; he has been running a monthly Seminar on molecular and physical gastronomy since 2000, and he gives the yearly Cours AgroParisTech lecture on molecular and physical gastronomy (public and free courses with a new course each year). In 2008, he was the organizer and president of EuroFoodChem XIV and he created the Groupe Français de Chimie des Aliments et du Goût of the Société

Française de Chimie. He is the French representative of the Société Française de Chimie at the Food Chemistry Division of EuCheMS.

Hervé this vo Kientza lectures extensively in all countries, and he contributes to the establishment of molecular and physical gastronomy laboratories in universities around the world. He contributes many columns in journals (in France and in other countries), and has published about 20 books, including: *Les Secrets de la Casserole, Révélations Gastronomiques, La Casserole Des Enfants, Science et Gastronomie, Casseroles et Éprouvettes, Traité Élémentaire de Cuisine, Six Lettres Gourmandes, Maths'6, Petits Propos Culinaires et Savants, La Cuisine, C'est de L'amour, De L'art, De la Technique, Construisons un Repas, De la Science aux Fourneaux, La Sagesse du Chimiste, Science, Technologie, Technique (Culinaires): Quelles Relations?, Les Précisions Culinaires, La Cuisine Note à Note, Mon Histoire de Cuisine*.

Hervé This vo Kientza is an honorary member of many academies, member of the Académie d'Agriculture de France, of the Académie de Stanislas, of the Académie Royale des Sciences, Des Arts et Des Lettres de Belgique, of the European Academy of Science, Arts and Letters, of the Académie d'Alsace Pour les Sciences, Lettres et Arts and he has received many awards and prizes, such as the Franqui Professorship (Belgium) and the Grand Prix des Sciences de l'Aliment by the International Association of Gastronomy. Recently he was nominated Doctor Honoris Causa by the University of Agricultural Sciences and Veterinary Medicine Cluj-Napoca (USAMVCN, Romania)

Hervé This vo Kientza has been knighted officer of the French Ordre des Arts et Lettres, commander of the Ordre du Mérite Agricole, officer in the Ordre des Palmes Académiques, and knight in the Ordre de la Légion d'Honneur.

CHAPTER 1

Let Us Stand on Our Two Legs

"BACK-OF-THE-ENVELOPE CALCULATION" is very popular among chemists or physicists. Amidst discussions in pubs, bars, cafeterias or restaurants, they would use any piece of paper that they can find, such as the back of used envelopes, for drawing chemical mechanisms, making short calculations, or testing assumptions. Such discussions are quantitative because chemists and physicists are scientists, and sciences of nature rely on numbers, on equations, as wrote Galileo Galilei:

Philosophy [of nature] is written in that great book whichever is before our eyes - I mean the universe - but we cannot understand it if we do not first learn the language and grasp the symbols in which it is written. The book is written in mathematical language, and the symbols are triangles, circles and other geometrical figures, without whose help it is impossible to comprehend a single word of it; without which one wanders in vain through a dark labyrinth.

(Galilei, 1623)

For describing this activity, French people have a different expression in which food is obviously present: "*calculs sur un coin de table*"; the translation is "calculation on the corner of a table". And usually this table should be in a restaurant, with a paper napkin on which you can write and calculate. For example, I remember well the French physicist Pierre-Gilles de Gennes, Nobel prize winner in 1887, doing so during lunches and dinners that I organized at the first International Workshop on Molecular and Physical Gastronomy (1992), in Erice, Sicily.

Also, our Group of Molecular Gastronomy (note: I speak of molecular gastronomy, that is physical chemistry; not of molecular cooking, that is cooking) goes once a week for lunch in a restaurant that is chosen because of the napkin (it has to be paper), in order to discuss scientific matters, ask questions, and solve problems. At one time, we had "daily questions" such as: why don't we fall through the floor (the title of a wonderful book by J. E. Gordon, 2006), or how much foam can one make from one egg white, or how many oil droplets in a mayonnaise sauce? Later, these "questions of

DOI: 10.1201/9781003298151-1

FIGURE 1.1 *Pierre-Gilles de Gennes (1932–2007) was a French physicist and Nobel prize winner in 1991 for his work on dispersed systems, among other topics.*

the day" were named "exercises", then "joggings", but the topic remained the same: because many students of our research group love cooking and science, we generally discuss questions of chemistry and physics with a relationship with cooking. Hence this book.

But calculation is not enough: it has to be based on experiments, as illustrated by a discussion that I had with Pierre-Gilles de Gennes (Figure 1.1) before the workshop that I mentioned above. During the lunch before the opening lectures, he asked me what we were going to do exactly. I answered that I was particularly interested in collecting and testing "culinary precisions", *i.e.*, culinary old wives tales, proverbs, and tips. Such pieces of technical information add to the "definition", the main corpus of the recipe, explaining how to reach the goal, *i.e.*, to make the dish. Pierre-Gilles wanted an example, and because on this particular day the napkin was a white tissue on which I could see bottles of wine, I answered by this culinary precision: it is (sometimes) said that when you make a stain of wine on the napkin, you can eliminate the stain by covering it with salt. Upon hearing my words, Pierre-Gilles immediately tried to interpret the effect, with diffusion, dissolution of salt, adsorbing phenols on salt, and so on. But I stopped him. First, I took the bottle of wine and poured some of the liquid on two places. Next, I waited a certain time (we counted), covering one of the stains with salt, but leaving the other untouched. Then, after some time (decided in common and measured), we compared the two stains and we agreed that they were the same.

Of course, we shared the idea that we have to interpret phenomena only when they exist really. But when the phenomena are demonstrated, many interpretations can sometimes compete, and then calculations are important in order to select the right ones.

Returning now from experiment to calculation, as such trips are frequent in science, we have to add that calculation can be symbolic, formal, or based on orders of magnitudes. Indeed, Pierre-Gilles de Gennes, after Richard Feynman (another Nobel prize winner in physics) and many others, was very good at "physics with the hands" because he was very good at calculating, avoiding bad explanations. And he was good at that—he told me once—, because he had been training himself intensely in calculating order of magnitudes.

He knew too well the story of Enrico Fermi—another Nobel prize winner in physics—asking students applying to his lab how many piano tuners there were in Chicago? One had to assume an order of magnitude of people in the city, an order of magnitude of pianos in homes, and an order of magnitude for the frequency of tuning, and then compare the number of tunings in a year by the number of possible tunings by one piano tuner in a year.

Pierre-Gilles was doing exercises daily of that kind of question. When walking in Paris to go to his lab, he asked how many stones to make the roads (Figure 1.2), how many buildings, how many tons of garbage per day, how many leaves on a tree at the corner of the street, how many litres of water flowing through the Seine river in one year, how many blades of grass in the Luxembourg garden (at the center of the Quartier Latin, where many top universities are located), and so on (De Gennes, 1998).

Now, the stage is ready: the "recipe" of this book is to mix experiments and simple calculation. But we still miss a clear vision of our goal: calculating and experimenting are not goals in themselves, but only the two feet on which we have to stand in order to make science or technology.

I know that some colleagues disagree (only some, and we can discuss), but I see a main difference between the two activities that are science (sciences of nature) and technology. Technology (from *techne*, to do, and *logos*, study) means improving technique, often with the results of sciences of nature. But for sciences of nature, the goal is to explore the mechanisms of phenomena, using the following method:

(1) identify clearly a phenomenon,

(2) characterize this phenomenon quantitatively (measuring all aspects of the phenomenon, focusing on the most pertinent ones),

(3) group—or synthesize—the data recovered in (2) in equations, also called "laws" (we shall discuss this term),

(4) induce a "theory", by introducing new concepts that are quantitatively compatible with all "laws",

(5) look for refutable consequences of the theories,

(6) test experimentally these "predictions", and so on back to (1), again and again, with no end.

FIGURE 1.2 *How many stones are there in a street? This is the kind of calculation that Pierre-Gilles de Gennes was doing daily. The training goes like this: let us assume that the street is as long as three Eiffel towers. This would be about 1 km, that is 10^3 m. If a stone has a side of 10 cm, or 0.1 m, and if the stones would be simply linearly aligned, this means that each row would have $\frac{1e3}{0.1}$, or 1e4 stones (elsewhere is the book, we use the symbol "e" for a power of ten; 1e3 means 10 to the power 3, or 1,000). Now, we need the width of the street: let us assume that four cars go be side by side, or about 8 m. This would make 80 stones. So that the number of stones in the street is about 800,000.*

This description of this wonderful enterprise that is science (of nature) clearly shows that in order to participate in scientific research, we need experimental and theoretical skills. As expressed in ancient Latin, *mens sana in corpore sano*, said the antique Latin people: a healthy mind in a healthy body, good experimental skills and good calculation skills. For technology also, both activities are important: theoretical ideas drive experimental activities, and experiments lead calculations and theoretical work.

Certainly, some students prefer experiments (often when they know implicitly that their calculation background is somehow weak), and other students prefer theory (when they are not trained to use their hands). But isn't it a pity to limit oneself to one activity only? This would be as if walking on only one leg! Why not stand instead firmly on two? Often, the "choice" is made as a result of a personal history, with only insufficient training leading to the conclusion that one is good at or bad at something. As students (I repeat that it means "anyone who studies"), we should not succumb to that, keeping in mind the other Latin proverb: *labor improbus omnia vincit*, hard work always succeeds.

And this precept determines the structure of this book: there will be both the practice of cooking (and experimenting), and calculations when the questions become clearer. The current appeal of cooking, with TV shows, online recipes, podcasts, etc. has rendered molecular and physical gastronomy fashionable (or is it the reverse: science has made cooking appealing to all, whereas more often it was girls who cooked in the past?). Whatever the cause, let us use this appeal for the sake of students, who are now willing to experiment and calculate. In all chapters, we shall begin by a recipe, and this experimental part will lead to calculations that seldom call for more than cross multiplication. Indeed, nothing should be very difficult: only a dozen of "laws" are needed.

Here, the word "laws" is in quotations, because, for some years, this term has been criticized by scientists (Ghose, 2013; Carroll, 2020). There is a difference between a law, in the sense of the civil community, and an equation that describes a set of experimental data, even if this equation then predicts the result of particular experiments that have not been done. For example, when the German physicist Georg Ohm (1789–1854) guessed some linearity between the values of electric potential difference and electric current intensity in a simple conductor, physics was able to describe the phenomenon by a simple equation, which was named "Ohm's law" (Ohm, 1827). Because the word "law" is not the same as a law that forbids killing or stealing, scientists increasingly use instead the word "models".

These models, or equations, or formulas, along with simple definitions of chemical and physical concepts, will be given and explained after an introductory chapter that discusses the experimental side of questions, well illustrating the phenomena. But giving the models and definitions is not enough, and a chapter will explain how to calculate, and how to deal with small problems (including inventing a particular calculation). Indeed, the strategy for calculation is probably more important than the calculations themselves.

The main body of the book is about asking small questions after an experiment has been proposed for making the questions clearer. Call these exercises, or problems, or "joggings" if you prefer. The questions are ordered from the simplest to the more complex, from gas, liquid, or solids to "colloids" such as gels, emulsions, foams, suspensions, aerosols and so on.

Here, it is important to declare that, for each question, there are many possible solutions. Do not be too shy. Propose the calculation that you want, based on your personal background, your personal knowledge, and your personal skills. This is because any calculation that you will propose is legitimate; after all, the fun is to invent one's own calculation, applying what one knows in order to investigate the particular experimental situation that is discussed. In the chapters that follow, one solution only (with exceptions, for validations) is given: it was selected after discussions with many students in our research group, and the level of complexity is always of the high school level.

REFERENCES

In the lists of references, for all chapters, some comments are sometimes made. . .

Carroll JW. 2020. Laws of nature, *Stanford Encyclopedia of Philosophy*, https://plato.stanford.edu/entries/laws-of-nature/, last access 2021-12-31.

The philosophical study of sciences of nature is called epistemology; and indeed, this encyclopedia answers to many questions that one can have about the nature of sciences of nature.

De Gennes PG. 1998. L'intelligence en physique, *Pour la Science*, 254, 14.

Yes, sometimes, you will find references to texts in French, but given the online translation systems that exist today, it will be easy to discover wonderfuls pieces such as this one.

Galilei G. 1623. *The Assayer*, https://web.stanford.edu/ jsabol/certainty/readings/Galileo-Assayer.pdf, last access 2021-12-31.

With Nicolaus Copernicus (1473-1543) and Francis Bacon (1561-1626), Galileo Galilei (1564-1642) was one creator of modern science. Do not hesitate to read his texts: they are so intelligent!

Ghose T. 2013. "Just a Theory": 7 Misused Science Words, *Scientific American*, https://www.scientificamerican.com/article/just-a-theory-7-misused-science-words/, last access 2021-12-31.

This piece is only one example of the epistemological debate about this important question of the "laws of nature".

Gordon JE. 2006. *The new science of strong materials: or why you don't fall through the floor*, Princeton University Press, Princeton, NJ.

I love questions such as the title of this book, as you would easily guess.

Ohm GS. 1827. *Die galvanische Kette, mathematisch bearbeitet*, TH. Riemann, Berlin, https://gallica.bnf.fr/ark:/12148/bpt6k33646.image, last access 2021-12-31.

It is always fascinating to read the works of the scientists of the past, in order to discover "methods".

Practical Skills

W HEREAS MANY students are interested in "food and cooking", as mentioned, we have to recognize that there is a major difference between cooking and physical chemistry.

Previously, we considered science (or more precisely sciences of nature), and we now know that it means looking for the mechanisms of phenomena using the method given in chapter 1. But cooking? Here, the goal is not to look for mechanisms of phenomena, but rather to produce food, or more precisely dishes. Why this difference between "food" and "dishes"? Food, according to the dictionary is any item that can be used for consumption, giving energy and building blocks to our body. Dishes, on the other hand, are ingredients culturally organized under a form that makes them "good".

More generally, cooking has three components (This and Gagnaire, 2010): (1) a technical component (when grilling meat, the colour should not be too dark, for example; when making an emulsion such as a sauce mayonnaise (Figure 2.1), the two phases—oil and water—should be correctly dispersed); (2) an artistic component (the question of reaching culinary beauty is not a technical one); (3) a social component (eating a dish that was prepared by someone requires trusting him or her).

The method for successful cooking should be deduced from this observation. Technically, the main issue is the "care": of course, the technical goal has to be reached (an emulsion must be emulsified, a foam must be foamed, etc.), but technicians implicitly or explicitly recognize also that the care is most important because, contributing to the technical success, it also demonstrates a social link between the cook and the guest. About art, there is much diversity, as in painting: although painting a wall requires care, it is not the same as painting as considered by Rembrandt, Dürer, Goya or Monet. The same kind of idea holds for cooking: the making of sandwiches is utilitarian, but some chefs create on another level. And finally, the most "beautiful" dish would be nothing if thrown in the face of the guests: the special relationship between the cook and the guest is most important.

All this being said, the difference between cooking and science—in particular physical chemistry—appears clearly: on one hand, the issue is exploring mechanisms; on the other hand, one has to make beautiful (*i.e.*, good) artifacts. Cooks can be

DOI: 10.1201/9781003298151-2

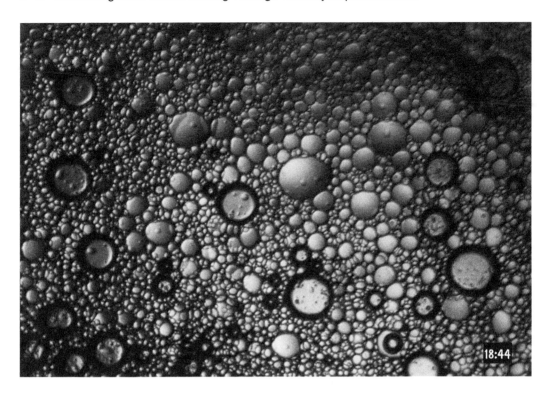

FIGURE 2.1 *A sauce mayonnaise is obtained by whipping oil in a mixture of egg yolk and vinegar (plus salt and pepper) (Carême, 1828). The energy given by the whisk divides the oil flow into microscopic droplets, which are dispersed in the aqueous solution made by the yolk (50% of it is water) and the vinegar (about 90%). This is an "emulsion", and perhaps one of the best example of such colloidal system.*

interested in science; they can be interested in observing the many phenomena that appear during the culinary practice, but it is not their goal to investigate them.

To conclude our discussion of the differences between sciences of nature and cooking, let us observe that too often, people confuse the word "science" with "knowledge"; in terms of knowledge, it is obvious that there is a "science of the cook", a "science of cooking". But this "science" is not to be confused with sciences of nature, such as chemistry or physics. Certainly, rigor is important both in the sciences of nature and good cooking, but rigourous, precise technique is not enough to make cooking a "scientific" activity. In summary, sciences of nature and cooking have very little in common: they have different goals, and, accordingly, different ways of reaching their goals.

Now that the meaning of "science" is clearer, let us examine the question of "experiments". In the past, cooks experimented little, and reproducing what was demonstrated by the professor or by an elder cook was the essence of culinary instruction. Happily, this has changed, and greater artistic freedom can be now attained.

In sciences of nature, on the other hand, experiments have been essential since the introduction of modern science during the Renaissance because they constitute the foundation of our theories. First experiments are important because you need to observe phenomena; here, science intersects with cooking, as testing culinary ideas can sometimes show new phenomena that can be later investigated scientifically (and this is indeed "molecular and physical gastronomy"). Galileo Galilei (1564–1642) expressed this:

A good way to reach the truth is to prefer experiment to any reasoning, because we are sure that, when a reasoning disagrees with experiment, it contains an error, at least hidden. Indeed it is not possible that a sensible experiment can be opposed to the truth. And this is a principle that Aristotle was praising, and whose strength and value are much over those that are based on the authority of any man in the world.

(Galilei, 1640)

Accordingly, experimenting well is of utmost importance for good science because how would you find a suitable equation if you do not have the right data from which to make it? What is "experimenting well"? This question is difficult and often poorly discussed, as demonstrated by many textbooks of "general chemistry" in which the first chapter is on safety, repeating "don't do this" and "don't do that". Too often, these introductory chapters are so long that the students skip them, or conclude that they do not know enough; therefore, they should not do anything. And because there is indeed a lot to know, students do not rank the information in order of importance, and they sometimes move without the right precautions and have accidents.

Should we not learn differently? For example, students are frequently advertised not to heat liquids in graduated glassware, but why? Would it not be better to invite them, on the contrary, to make "graduations" on a simple and cheap glass tube, to seal it at the bottom, to fill it in with water, and to heat it, observing that the tube breaks where it was graduated? During this experiment, without dangerous liquid, students would learn to make a graduation (almost no risk), to seal a tube (a bit of risk, but useful for chemistry), to fill a tube with a liquid (no risk, useful), to heat a tube containing a liquid (no risk)... and to understand that graduations weaken the glass, which leads to breaking when heated.

More generally, would it not be better to propose explanations instead of rules? Also, students are told to put on glasses and gloves in the lab and this is a very important safety practice indeed, but the issue is not to say something, but rather to demonstrate it: why not inviting them to add a colored liquid in a beaker in which a magnetic stirrer is moving, a piece of white paper being put under the beaker? Sooner or later, they will see tiny drops of colored solution on the paper, demonstrating that there were some projections: the students would then understand that the same contamination occurs with toxic colourless liquids. In chemistry, the danger is greater because it is hidden, and coats and glasses are important because we do not wish to die, do we?

With cooking, the risk is less, even if heat and knives are used, but the same precautions are important, as demonstrated by the use of a pastry bag, or even simply cutting a brunoise (small dice) from a carrot. In the section 6.6 of chapter 6, a contest involving filling a glass over the edge is proposed: clearly, the intelligence

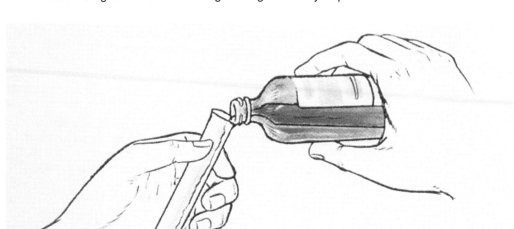

FIGURE 2.2 *There has been a lot of progress in safety in the chemistry laboratory (here the chemist is not wearing gloves!). On the other hand, the old chemistry books gave useful experimental advice, and good chemists were recognized by the application of this advice.*

of the hand can work in concert with the one of the head. And this is why students, in proposing their experiments, are invited to ask themselves about the "how to".

Sometimes the physical or chemical analysis of the manipulations to be performed is enough to get an idea of how to perform them well, but sometimes more knowledge is needed. Therefore, one would be well advised to ask oneself the question "What is the question I don't think of"?. This is one of the most inviting questions one may ask that avoids having us fall into the trap of straightforward activity without reflection, the one that we master when moving the hands, so that the hands become as smart as the head.

I am not suggesting that all introductory chemistry books are bad, but one has often to look into old chemistry books in order to get these useful advice which belong to the lore of chemistry and that is often transmitted only orally in laboratories (Figure 2.2).

How to open a bottle without putting the cork on the bench? How to pour a liquid so that it flows exactly where one wants? How to transfer a powder? How to curve a glass tube? Regarding these questions, let me share with you my enthusiasm about the wonderful book *Chemical Manipulations: Being Instructions to Students in Chemistry, on the Methods of Performing Experiments of Demonstration or of Research with Accuracy*. It was written by Michael Faraday, the "prince of lecturers", but also the "prince of experimenters". He owed this last skill to his apprenticeship as a book binder, and no one could have better written a book on chemical manipulations (Faraday, 1827). In the foreword, Faraday writes:

Chemistry is necessarily an experimental science: its conclusions are drawn from data, and its principles supported by evidence derived from facts. A constant appeal to facts, therefore, is necessary; and yet so small, comparatively, is the number of these presented naturally to us, that, were we to bound our knowledge by them, it would have but a small extent, and in that limited state be exceedingly uncertain in its nature. To supply the deficiency, new facts have been and are created by experiment, the contrivance and hand of the philosopher being employed in their production and variation. In reference to the varieties of inert matter, their possible forms, states, and properties, and the powers which influence them, Chemistry, if occupied only in the observation of such phenomena as are presented by Nature, would do little more than record the comparatively quiescent state of things which has followed the active exertion of the inherent powers of matter; and the chemist would have but little opportunity of observing substances in their energetic state, or of witnessing the phenomena which these powers are able to produce.

Yes, experimenting is not simply following a protocol, but rather putting intelligence into what we do, being always curious of all that does not behave as we think that it "should". Some speak of "serendipity" in relation to the great scientists of the past, but does serendipity mean more than simply being attentive to all that happens, being surprised by everything, continuously looking for oddities?

Certainly, this is something important that we have to keep in mind when performing the experimentation proposed in the following sections. The hands with the head, always the hands with the head, and also the head with the hands!

REFERENCES

Carême A. 1828. *Le Cuisinier parisien, ou l'art de la cuisine française au 19e siècle*, Chez l'auteur, Paris, France.

Marie Antoine Carême (1784–1833) is probably the most famous chef of the history of French cuisine. He was the cook of kings and emperors, and this book is a cornerstone of haute cuisine. His book was translated in 1836 under the title French cookery: comprising L'art de la cuisine française, Le patissier royal, Le cuisinier parisien, see https://archive.org/details/b29338098, last access 2022-01-10.

Faraday M. 1827. *Chemical manipulations: being instructions to students in chemistry, on the methods of performing experiments of demonstration or of research with accuracy*, John Philipps, London, UK.

I cannot emphasize this enough: Faraday was a wonderful scientist, not only from the scientific point of view (he discovered benzene, electromagnetic induction, the Faraday effect, the equations of electrolysis, diamagnetism, and the line of forces, he liquified several gases, he was the founder of nanoscience and more... I have to stop here because he did too much to be adequately represented in such a small comment.

Galilei G. 1640. *Letters to Fortunion Liceti*, https://www.researchgate.net/publication/262592609_Letter_from_Galileo_Galilei_to_Fortunio_Liceti, last access 2022-01-11.

This H, Gagnaire P. 2010. *Cooking, a quintessential art*, University of California Press, Berkeley California, USA.

The French title of this book of mine (containing recipes by Pierre Gagnaire, one of the best chefs in the world) was "La cuisine, c'est de l'amour, de l'art, de la technique".

CHAPTER 3

Theoretical Skills

T HE PREVIOUS CHAPTER discussed experimenting, with the hands and the head. The head? It is not enough to think well, almost by luck, in the sense of drawing logical consequences from premises; we would benefit from having tools for that. And these tools are called notions, concepts such as energy, chemical potential, entropy, surface tension, forces, differential and integral calculus.

Students in chemistry, biology, or food science and technology are sometimes less prone to using these tools than students in physics, and many admit that once exams are finished, they forget what they learned. Later, many have retained only a vague understanding of such topics, and they justify this by saying that they would learn them again if needed.

However, it is a fact that many will certainly need calculation skills for their future careers. Since modern science and technology is increasingly based on calculation, calculation skills have to be developed during education (Owenson *et al.*, 2012; Feser *et al.*, 2013). Thus, one goal of this book is to help students to improve their calculating skills, to gain confidence in themselves with easy questions, and to recognize that it does not require much effort to transform knowledge into skills. Culinary practice is useful because processes performed in the kitchen go along with physical or chemical changes that can be investigated quantitatively using simple physical and chemical knowledge. In other words, asking questions about cooking is a good way to invite you to improve your calculation skills.

Sometimes the rejection of math is at the root of weakness in calculation. Some students do not understand why the mathematics is interesting, and they would prefer to see how the mathematical concepts will be useful before learning them. In this book, this will appear clearly.

But the discussion is not so simple, as clearly stated by George Polya, in the Introduction to his wonderful *How to solve it*:

A great discovery solves a great problem, but there is a grain of discovery in the solution of any problem. Your problem may be modest, but if it challenges your curiosity and brings into play your inventive faculties, and if you solve it by your own means, you may experience the tension and enjoy the triumph of discovery. Such experiences at a susceptible age may create a taste for mental work and leave their imprint on mind and character for a lifetime.

DOI: 10.1201/9781003298151-3

> Thus, a teacher of mathematics has a great opportunity. If he fills his allotted time with drilling his students in routine operations, he kills their interests, hampers their intellectual development, and misuses his opportunity. But if he challenges the curiosity of his students by setting them problems proportionate to their knowledge, and helps them to solve their problems with stimulating questions, he may give them a taste for, and some means of, independent thinking.
>
> Also, a student whose college curriculum includes some mathematics has a singular opportunity. This opportunity is lost, of course, if he regards mathematics as a subject in which he as to earn so and so much credit, and which he should forget after the final examination as quickly as possible.
>
> (Polya, 1945)

Of course, the last sentence, including the word "mathematics", can be disputed, because mathematics are... mathematics, whereas calculations means using mathematical results for scientific (in particular) purposes, but nonetheless, the question of solving problems is important. Without such skills, how could we investigate quantitatively the phenomena that are studied?

REFERENCES

Feser J, Vasaly H, Herrera J. 2013. On the Edge of Mathematics and Biology Integration: Improving Quantitative Skills in Undergraduate Biology Education, *CBE-Life Sciences Education*, 12, 124–128.

Owenson S, Heath M, David B, Sleath C, Lynne I. 2012. *Maths and biology skills for life*, https://bbsrc.ukri.org/documents/1204-maths-in-school-level-biology-pdf/, last access 2020-05-04.

Polya G. 1945. *How to solve it*, Princeton University Press, Princeton, NJ.

One of those book that I treasure! So smart.

CHAPTER 4

Calculation

For Some of Us, It Is Fun; for Some Others, It Is a Necessity. It Can Become Fun after Realizing That It Is Easy

4.1 SIMPLE KNOWLEDGE

In this book, we selected questions that can be solved using scientific knowledge at the bachelor level in theory. However, in practice, many students at the master level were not able to find the solutions to some questions, because of insufficient training in solving them. Let us say it differently: studying these questions does not call for an important theoretical background. But a good approach and sometimes a calculation strategy are needed. In particular, determinations of orders of magnitude or "calculations on lattices" (this will be explained below) are extensively used.

More precisely, most questions of this book do not call for much more than the simple following models (remember our discussion about "laws") given below. Some of them are definitions, and others are very basic equations, formulas, or relationships of physics. Of course, some knowledge of chemistry is also useful: it will be found online when needed (chemical nature of cellulose, of various sugars, of triglycerides, of proteins...). At any rate, without delay, let us turn to the important bridge between chemistry and physics shown in Figure 4.1 (Cottrell 1959; Benson, 1965).

4.1.1 Handful of Simple Equations

I said earlier that a dozen formulas will be enough for discussing quantitatively the questions proposed in the main body of this book. We shall consider them now.

First, we have to say that the goal is not to make a full course of mechanics, thermodynamics, chemistry, and physics. You have done. This before, and you will material this before, and you will have to go back to your various courses if you want to see how these formulas were derived, and under which assumptions.

Sometimes they are simply definitions, and sometimes they are the result of more elaborated, theoretical, reasoning (not given). I urge you to relearn them if you have forgotten them... so you can remember them for the rest of your life. Because they are

DOI: 10.1201/9781003298151-4

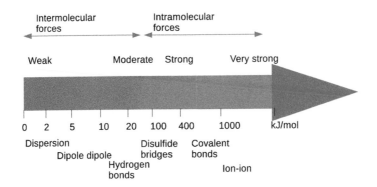

FIGURE 4.1 *The energy of various bonds between molecules or ions is given here. The chemical details of the interacting objects can change the energy of the bonds between them, it is important nonetheless to remember the order of magnitudes when analyzing a question.*

constantly essential in science or technology (of food). Or, to put it another way, if you know them, you will be able to find ways to apply them in the various circumstances of your daily life so that the world becomes more understandable.

These definitions and models are mainly about "physics", but I repeat that elementary knowledge of chemistry is certainly more than useful: the molecular structure of ethanol, the formula of acetic acid, the means by which acetic acid dissociates in water, equilibrating the chemical description of a chemical reaction, etc. can be added in passing. This information will be given and discussed later each time it is needed. Certainly, the chemical meanings are as useful as the physical ones. Why should we stand on one foot only?

Now for the definitions and equations:

1. The density ϱ of a body B is defined as:

$$\varrho = \frac{m}{v}$$

where m is the mass and v the volume (PAC, 1996).

Indeed, even if this notion is introduced in high school, the difference between density and weight remains sometimes confused; this is sometimes due to the confusion of "light" (resp. heavy) and "less dense" (resp. more dense), or because it is faster to use the terms "light" and "heavy". Also, when asked why oil floats on water, some students will explain that it is "because oil is less dense" than water, which is obviously a tautology, and not an explanation. All this shows that culinary questions using density serve as good opportunities for useful exercises; of particular use is the example of cocktails with many layers (see Chapter 9).

Finally, let us observe that this definition provides an opportunity to observe that density is a useful intellectual tool (as is the hammer for the blacksmith), creating new ideas, and describing the world better than can be done by usual words.

Here, I also have to say that I am sad for non-French students because they have to know this definition instead of finding it from words: what is called "density" in English is more explicitly called "masse volumique" in French. And "masse volumique is a synonym of "mass per volume". So you see that by having the words, you also have immediately the formal expression, and you divide the mass by the volume.

At any rate, understanding what we want to characterize quantitatively is certainly an even better solution than relying on words: the idea is to know how much quantity of matter (that is mass) is in a certain volume. If there is much mass for a certain volume, the object is dense, and its density is higher.

A preliminary detail should be noted here if you don't know it yet: in scientific books there is an agreement that any letter that represent a quantity is set in italics; and, as you can see, ϱ, m and v are here written in this way, in contrast to the representation of B, the letter that was used to designate a particular body (it was a name, not a quantity).

2. Pressure P:

$$P = \frac{F}{A}$$

where F is the force applied on a surface of area A (PAC, 1996).

Here, also, the definition is the simplest that one can find; when considering a force acting on a surface, the more area, the more distributed the force, and the less force per area.

Let us insist because this is useful advice: we think better when using the right words (Lavoisier, 1790). Here, it is important to make the difference between the surface and its area: the surface is the surface of the object being considered, and the area is a quantitative measurement of the extension of the surface. This is by no means a useless detail because the proper use of words often leads to proper calculations. . . . formal calculation is based on natural language.

3. The area and the volume of a sphere (or rather a ball, because the sphere is the surface of the ball) are respectively $4\pi r^2$ and

$$\frac{4}{3}\pi r^3$$

where r is the radius.

Concerning these expressions, it is worth telling the students that the great Greek mathematician Euclid (about 300 BC) knew only the cubic relationship between the radius and the volume of the inside of a sphere (a ball), and Archimedes (about 220 BC) was so proud of having found the expression for the volume of a ball that he asked that his tomb would show the picture of

a sphere is two third of the volume of the circumscribed cylinder. Archimedes was the one to prove that the volume of the cylinder is twice the volume of the sphere (Weisstein, 2022).

4. The kinetic energy E_c of a moving point object can be expressed by:

$$E_c = \frac{1}{2}mv^2$$

where m is the mass of the body, and v its velocity.

Of course, this expression holds only in certain circumstances (translation, no rotation, for example), and only for classical, non-relativistic mechanics, *i.e.*, at speeds well below the speed of light, but let us begin simply and keep this simple definition.

Here, one should add that energy is of course a most important concept for the analysis of physical questions.

5. The mechanical work W of a force of intensity F acting on a body that moves by a distance x in the direction of the force is:

$$W = F \cdot x$$

Here this expression is valid only for a force parallel to the displacement, and only if the force is of constant intensity during the displacement. When the force is changing with position or with time, one has to use the "differential" form:

$$dW = F \cdot dx$$

where d stands for an infinitesimal variation.

All students are not always fully comfortable with differential and integral calculus. However, this can be solved easily with the wonderful *Differential and Integral Calculus*, by Nikolas Piskounov (1996). Students should read it slowly (word by word, line by line), completing all exercises of the book one by one. Also, they can get more comfortable if they write instead:

$$\Delta W \approx F \cdot \Delta x$$

as rigorous equality holds when Δx tends toward 0.

Also, it should be added that a simple way to take into account the collinearity of the force and the displacement is to write the formula with vectors.

6. Newton's second equation of dynamics:

$$\Sigma \overrightarrow{external\ forces} = m\vec{a}$$

The first equation of motion, also called the "equation of inertia", states that an object at rest in a certain "Galilean" referential remains at rest, and an object that is moving will continue to move straight and with constant velocity if, and only if, no net force acts on that object. For example, in a train moving at constant velocity, in a straight line, a ball on the floor will remain still.

And the second equation applies in particular to a body upon which external forces (think of a balloon being kicked) are applied.

Here m is the mass of the body on which the external forces act, and a is the acceleration of the body, that is the first derivative of the velocity v with respect to the time t, or the second derivative of the distance with respect to the time.

In more depth, it is perhaps better to remember this equation as:

$$\Sigma \ \overrightarrow{external \ forces} = \frac{\mathrm{d}}{\mathrm{d}t} \overrightarrow{p}$$

where $\overrightarrow{p} = m\overrightarrow{v}$ is the "momentum" because this second form is more general, applying also to cases where the mass changes with time, as when you throw away stones behind the boat on which you are floating, or when a rocket ejects gases.

7. The model (or formula) of ideal gases:

$$PV = nRT$$

where P is the pressure of the gas, V its volume, n the number of moles, R the constant of ideal gases $(8.314 \ \mathrm{J} \cdot \mathrm{mol}^{-1} \cdot \mathrm{K}^{-1})$, and T the absolute temperature.

Of course, this "equation of state" applies only to gases... to which the equation applies, but I invite you to look for information about how real gas is different from the ideal one: this is how one learns more and more.

This equation will allow you to calculate the following very useful information: the vaporization of one mole of any compound makes a volume of 22.4 L (at the temperature of 0°C and at the atmospheric pressure 101,325 pascals (Pa)) or 24 L at the temperature 20°C, always at standard pressure), but this is equivalent. (One question in this book invites the students to demonstrate it.)

8. The expression of the electric force F between on a body with electric charge q_2 from an electric charge q_1:

$$F = \frac{1}{4\pi\varepsilon_0} \frac{q_1 q_2}{r^2}$$

where ε_0 is the permittivity of the vacuum (farad, F, by meter, m) and r the distance between the charged bodies (in m). The force is in newtons (N) when the charges are in coulombs (C).

Do you wonder where this formula could apply? Remember that food is made of organic compounds, and that some can be carboxylic acids, for example, with a possibility of having a negative charge (when deprotonated) that can attract positively charged ions such as sodium or calcium.

Of course, electricity is only part of the electromagnetic game. Other equations for magnetism could be introduced...but we wanted to limit the list as much as possible. Let us also observe that our list does not permit students to avoid learning dynamics, electricity, magnetism, thermodynamics, optics, chemistry of solutions, organic chemistry, etc. This book cannot replace courses the cover such material but only provide them with the confidence that with the acquisition of some easy knowledge, they can analyze practical questions.

9. For some questions, the Stefan equation (also called the Stefan-Boltzmann equation) can be useful:

$$P = A\sigma T^4$$

Here P is the power (quantity of energy per unit of time) radiated by a "black body" across all wavelengths, A is the area of the emitting body, σ is the Stefan-Boltzmann constant, and T is the absolute temperature (Stefan, 1879).

When the emitting body is not black, but "grey", *i.e.*, when it does not absorb all incident radiation, it emits less total energy, and this can be characterized by the emissivity ε, with the modified formula:

$$P = A\varepsilon\sigma T^4$$

About this equation, I invite you to be surprised (amazed?) that we find a power of 4, while most equations in physics have exponents 1 (Ohm's relation, expression of the weight according to the mass, etc.) or 2 (decrease of the gravitational force or of the electric forces).

10. The heat (energy E, sometimes noted Q) needed for heating a body of mass m and of heat capacity c_p by a temperature difference ΔT is:

$$E = mc_p\Delta T$$

About this equation, I want to insist on the E/Q question: some students are trained to use one particular symbol for a particular quantity (for example the energy, or the molar mass, or the area), and they find it difficult to change. If I were them, I would train in this regard, because intelligence also means adaptability to changing circumstances, doesn't it?

11. Sometimes, the Fourier equation for heat transfers in stationary conditions is needed:

$$\varphi = -\lambda\frac{\Delta T}{\Delta x}$$

where φ is the density of thermal flux, λ the thermal conductivity, T the absolute temperature, x the distance. Here, the equation is given for a one-dimensional transfer, but the increase in complexity would not be great with three dimensions.

When the temperature varies with time, the equation is (again for one dimension):

$$\frac{\partial}{\partial t}T(x,t) = \frac{c_p}{\varrho c}\frac{\partial^2}{\partial x^2}T(x,t)$$

with ϱ the density, and c_p the heat capacity.

The variations of concentrations of the above obey the same equation, but bear the name of the German physiologist Adolf Eugen Fick (1829–1901), who in turn made them after the French mathematician Joseph Fourier (1768–1830) (Fourier, 1822; Fick, 1855).

12. The "entropy" can certainly be defined through the ratio of the variation of heat by the absolute temperature, in reversible circumstances, but it is often appropriate to use the definition in terms of statistical physics:

$$S = k_B \ln(\Omega)$$

where k_B is the Boltzmann constant ($\sim 1.38 \times 10^{-23}$ J $\cdot$ K^{-1}) and Ω is the number of accessible microscopic configurations that are consistent with the macroscopic quantities that characterize the system (such as its volume, pressure and temperature). Of course, this rule holds only in some circumstances, but let us begin simply (Reif, 1967).

Finally, do you know that this wonderful formula is written on the grave of Ludwig Boltzmann, in the Central Vienna Cemetery (Wikipedia, 2022)? Certainly, it is a wonderful formula!

13. The surface tension γ is defined as:

$$\gamma = \frac{\partial}{\partial a}F$$

where F is the Helmholtz energy (free energy), and a the area (you may use A if you prefer, to have the same symbol as in the definition of the pressure, but trust me, it is the same).

This notion of surface tension is obviously important for food because most food systems are of a colloidal nature, such as emulsions (dispersion of oil in water, for example), or foams (air bubbles in liquids), for instance: there are surfaces, or even interfaces, between the various phases.

Colloidal? Colloids? When one wants to know what a chemical term means, it is good to look it up in the Gold Book from the International Union for Pure and Applied Chemistry (IUPAC, 2019). Here, one finds: "*The term refers to a state of subdivision, implying that the molecules or polymolecular particles dispersed in a medium have at least in one direction a dimension roughly between 1 nm and 1 μm, or that in a system discontinuities are found at distances of that order.*"

This means that the area of the interface between two non-miscible phases (oil and water, or air and water) increases much during the dispersion, implying that a lot of (surface) energy is needed to make it.

14. The Maxwell-Boltzmann distribution, giving, at the thermal equilibrium, the probability of a state of energy E:

$$p(E) = K \exp\left(-\frac{E}{k_B T}\right)$$

where K is a constant that one can get by "normalization" (the sum of the probabilities of all possibilities is equal to 1) on all possible states, k_B the Boltzmann constant and T the absolute temperature.

Assuredly, this equation is not the simplest, and we shall see that using it is not easy. But let us have some ambition.

We referred to initially "a dozen definitions and models" but there are actually fourteen...the exact number of "commandments for cooking" that I give now (This, 2014): these ideas are probably the most important ones to keep in mind when cooking, from a technical point of view.

1. Salt dissolves into water.

   Certainly in the case of table salt, there is a solubility limit, about 360 g per liter at 20°C, but when complex colloidal systems such as food ingredients and food preparations are considered, it is good to remember that salt can dissolve in their aqueous compartments (Lide, 2005).

   Here and below, "salt" refers to the usual more or less pure sodium chloride that we eat daily, but this commandment is more general: the same idea holds for sucrose (table sugar) with a different solubility (about 2 kg per liter at the temperature of 25°C), and other water soluble compounds that are used in the kitchen (tartaric acid, citric acid, etc.).

2. Salt does not dissolve into oil.

   Yes, even after centuries, sodium chloride would not dissociate into sodium ions and chloride ions in oil and it would not dissolve. The same idea holds for many compounds that dissolve in water: sucrose, tartaric acid, etc. This property can be used in particular when one wants to protect salt (and other similar compounds) from dissolution: wrap them in "oil".

3. Oil does not dissolve in water.

   By "oil", food scientists and food technologists mean a liquid mixture of many triglycerides, and, certainly, the solubility of triglycerides in water is known to be very low.

   But wait! Above, the appearance of an adjective and an adverb makes me immediately think to alert my friends: in science, adjectives and adverbs words should be replaced by the question "how much?". A way of characterizing quantitatively the distribution of a compound between water and a non-polar solvent (such as n-octanol) is to use a measurement called the $\text{Log}P$, *i.e.*, the logarithm base 10 of the partition coefficient between n-octanol and water (NCBI, 2022).

For example, for tristearin (IUPAC name 2,3-di(octadecanoyloxy)propyl octadecanoate), one can estimate the $\text{Log}P$ to be equal to 25.2: this means that the compound would be more than 10^{25} times more soluble in n-octanol than in water. This is only a theoretical estimation but triglycerides, like many odorant compounds that can dissolve in oil, have very small solubility in water.

4. Water evaporates at any temperature, but it boils at 100°C.

Leave a glass of water in the open air, and it will be empty after some time; even if the temperature is room temperature, *i.e.*, much less than the boiling temperature: 100°C at the level of the sea. Evaporation can be important in the kitchen, in particular for hot emulsions such as certain sauces, because when the water content is less than 5%, the emulsion "fails", with a phase separation.

One can add to this commandment that the vapor pressure increases with temperature until 1 atmosphere at 100°C, and that the boiling temperature depends on the pressure: water can boil at room temperature in rotary evaporators of chemistry laboratories, and it also boils at less than 100°C in the mountains (the boiling temperature is reduced to 85°C only over 4500 m), but the boiling temperature is increased in pressure cookers (Lide, 2005).

5. Most often, fresh products are made primarily from water (or another fluid).

"Fresh products" often means meat, fish, vegetables, or fruits. And some order of magnitudes can be usefully remembered (Ciqual, 2022). For example, meat is about 70% water, and 20% proteins. Or plant tissues typically contain 80 to 99% water. Yes, salad is 99% water, for example, and I note this because isn't it wonderful that a sheet of lettuce, that does not "flow", contains as little as 1% of compounds to make it "solid"? Obviously, this chemical and physical system is well organized. . .

6. Foods without water or another fluid are hard.

Put a slice of meat or of carrot in an oven at 95°C for a long time (> 12 hrs) and you will be sure of this commandment. After all, water evaporates quite quickly at this temperature, and all other non-evaporable compounds remain (proteins, triglycerides, sugars and other saccharides, etc.). If the content in triglycerides is low (remember: we said "or another fluid"), then the dried material is solid.

7. Some food ingredients with high content in certain proteins (eggs, meat, fish) coagulate.

For these proteins, there are various categories, such as egg proteins (they coagulate when heated), gelatin (it does not coagulate when heated, but it can make a gel when a solution of it is cooled), caseins from milk (milk can coagulate when it is added with acid, or rennet), some proteins from milk serum (they coagulate with heat), etc. (Belitz *et al.*, 2004).

8. The collagenic tissue dissolves in water when the temperature is higher than 55°C (Kopp *et al.*, 1977; Goutefongea *et al.*, 1995; Duconseille *et al.*, 2015).

The collagenic tissues wraps the muscular cells that are called fibers; it groups the fibers in bundles, and the bundles in bundles of bundles. When heated in water, the tissue is first disorganized ("denaturation"), leading to contraction of the meat, before dissolving in water and being partially dissociated: this is how gelatine is formed.

9. Foods are generally dispersed systems.

Dispersed systems? This means gels, suspensions, emulsions, foams, etc. Remember the definition of colloids given above.

10. Some reactions (glycation, oxidations, caramelizations, pyrolysis...) generate new compounds.

A simple example is sucrose (IUPAC name (2R,3R,4S,5S,6R)-2-[(2S,3S,4S,5R)-3,4-dihydroxy-2,5-bis(hydroxymethyl)oxolan-2-yl]oxy-6-(hydroxymethyl)oxane-3,4,5-triol) that turns brown when heated. A lot of new compounds are produced, and all do not have a color or an odor, but some have. The number of possibilities is really infinite because food ingredients and the conditions of their processing are so diverse!

11. When food becomes whiter, during a process, it is often the result of some foaming or emulsification, or coagulation.

Of course, this holds only if the light is white, because the appearance results from the reflection of light on the surface of the objects that make up the system. For example, a foam is white, under white light, because the light reflects on the walls of the bubbles; but a whipped egg white (foam) illuminated with red light is red.

12. Capillarity moves liquids.

And this occurs each time a liquid is near a slit, a crevace, or a tiny hole, for example (De Gennes *et al.*, 2003).

13. Osmosis.

This process takes place when liquids having different concentrations are separated by an appropriate membrane. This membrane is said semipermeable (This vo Kientza, 2021a).

14. Compounds can move through diffusion.

About "diffusion" there is much confusion, and many people use the word erroneously each time compounds move during cooking (Aguilera *et al.*, 2004). But capillarity is not diffusion, for example. Suffice it to say here that a molecule in the middle of a liquid can "diffuse", *i.e.,* move randomly, either because of its proper thermal motion or because of interactions with the molecules of the liquid.

Now an equilibrium is found between science and cooking, isn't it?

4.1.2 How to Apply the Equations

Having these equations at hand is not enough: we have to use them. But how?

In the context of this book, students are invited, when tackling a question, to simply consider the 14 definitions and equations one after the other, and to ask themselves if they apply in the particular circumstances that they meet. If yes, all the best; if no, the next equation is to be considered.

For example, for a question with no motion, the kinetic energy is probably useless. And when there is no radiation, the Stefan-Boltzmann model can be put aside.

But this advice is not enough, and students will be better advised to analyze the questions first, in order to select the right physical or chemical concepts from their scientific background.

This can be quite a challenge for some, as exercises and problems of physics and chemistry from their curriculum were often limited to applying freshly learnt formulas in a restricted calculation environment; *e.g.*, while studying electricity, students would have to use the expression of the electrostatic force with particular data.

Here, on the contrary, they will have to first pose the problem clearly and make assumptions, which is sometimes difficult: for instance, many years of teaching at the master level showed me that very few students could calculate the thickness of the crust of a soufflé (see Chapter 11, Section 11.2).

4.2 HOW TO CALCULATE

Before providing the questions with one or two solutions for each, let us analyze why some students find calculations difficult.

First, some of them do not know that calculation and natural language are almost the same. In the past, symbolic calculation was introduced in order to simplify mind processes (Jongsma, 1979). The Welsh physician Robert Recorde (1512–1558) introduced the signs "=" and he popularized the sign "+". Shortly after, the French mathematician François Viète (1540–1603) developed the idea of representing variables and constants by letters. Then René Descartes (1596–1650), Gottfried Wilhelm von Leibniz (1646–1716) and others improved the language of symbolic algebra.

Today, thanks to these beautiful minds, writing "$\frac{d}{dx} f(x) = \frac{\cos(x)}{x^2}$" is easier and faster than writing "the derivative of the function f of the real variable x relatively to this variable x is equal to the ratio of the cosine of the variable x by the square of x" (one does not remember the beginning of the sentence when it is finished!).

Unfortunately, many students do not know that this symbolic notation is to help them, and they consider instead that their life is made more complex by incomprehensible symbols. We have a duty to tell them that formal calculation is easy, simple, and powerful, and it can even work quite automatically (even if it is better to understand what one is doing).

Moreover, it helps students to know that because formal calculation is based on natural language, calculating means speaking and thinking as usual (but well), using the natural language. One can think of a formula as one would think in natural language and the same holds for algorithms and computer programs.

A first conclusion can be given: in order to calculate well, one has to think properly and to speak properly (This, 2017; 2021).

Of course, speaking properly, but with logic, can be difficult for some who "think in all directions", being disturbed by the many perturbations of life: having enough money for food or lodging, being in love, organizing activities, answering emails, and so forth. However, calculation cannot be done chaotically: it is an activity that goes rationally, logically, toward its goal.

One way to help students is to explain that if a calculation is like a story, one should divide it into:

- a starting point

- a path

- a goal.

This simple analysis has consequences:

1. if a student does not know where he or she is going (the goal), from the calculation point of view, the probability of reaching the goal is nil. To take a comparison, starting from Paris, the probability of reaching Berlin is nil if one does not know that he or she wants to go to Berlin.

2. if it is not known which way will be used to go from the point of departure to the point of arrival, there are few chances again that the goal will be reached. In other words, a way is needed. About that, let us remember that in Greek, the way is called *methodon*, the word for "research of a method". Indeed, often students can find the goal, but they have difficulties in finding the way. They can be helped by being invited to look first to a "strategy" (the general idea) and only later to a "tactic", *i.e.*, the practical way to move.

For example, when it has been decided to travel from Paris to Berlin, one has to decide to go either by plane, or by train, or by car, or by bicycle. When the train (for instance) is chosen, and only then, one can decide to walk from home to the bus station, then to take the bus to reach the railway station, then into the train, then, when the train has arrived in Berlin station, to go down the to train, then to take a cab to reach the precise final destination.

Indeed, we are fortunate that for many small problems in this book the solution is obvious when the starting point is well defined and when the problem is clearly stated. Often it is enough to translate each sentence of the statement, in natural language, into a formal expression or a computer code, depending on the circumstances. This method has the advantage that the final calculation will be documented; one will be able to check it, and any reader (including the professor checking it) will be able to follow.

Finally, we propose to invite the students to always structure their argument with:

1. the question

2. the way to reach the goal, including the choice between a simple use of orders of magnitude for simple questions, or formal writing for more complex problems

3. the particular steps

4. the recognition of having reached the goal.

However, the practice of teaching shows that failures can occur even when the general method is good. The reasons for failure are frequently the following:

- mixing formal calculation with numerical calculation,

- using units that do not belong to the International System (did you ever visit the official site?) (BIPM, 2020),

- lack of structure of the document,

- poor marking of the formal symbols of figures, leading to confusion (between 9 and g, or 1 and 7, for instance),

- bad translation of an idea into an equation; often, this means that the calculation does not include text in natural language;

- jumps in the algebraic manipulation of equations (instead of step-by-step changes).

But let us be positive and let us now examine the various parts of any calculation (This vo Kientza, 2022):

1. *The question considered.* In most exercises, the question is simple to formulate and it is easy to translate natural language into formal language.

   In some problems (when they are not a succession of exercises), one has to learn to ask questions to oneself, and to create a way from the initial data to the final question. But here again, students have to learn the easy step of introducing a formal symbol for any unknown quantity. More generally, they have to get into the habit of characterizing formally any "object" by one or many symbols.

2. *The way to solve the problem.* Often, using order of magnitudes can be useful and fast, as explained in the introductory chapter. And when such fast calculation is not possible, one has to move to formal calculation. This is where software for formal calculation such as *Maple* (Maplesoft, 2022) are useful, and the possibility of dividing the document into sections and subsections is particularly helpful because it structures the work. Of course, sometimes there is also the possibility of making a digital model.

3. *The step-by-step way.* Here it is often needed to "put down ideas on paper" (in the past, it was paper, but now computers are preferable) so that one can analyze the words.

This should be done in natural language, and it is only when the job is finished that the formal symbols are introduced.

The structure should be:

- Analysis of the question
- Expressing the result
- Numerical application.

In the analysis step, looking for relationships between the already introduced symbols is often enough to find the solution.

But let us come back now to the "analysis of the question". This should be divided into:

- Writing down the parameters (from the question),
- Making a qualitative model; this is frequently a synonym of making a picture, and analyzing the question physically or chemically. Here is the place for simplifications and assumptions,
- Turning the qualitative model into a quantitative model through the introduction of formal symbols for describing the characteristics of the model. If a picture is drawn, this is very easy, because one has simply to put letters for each aspect of the various parts that one can see on the picture, such as shapes, colors, surfaces, etc.
- Looking for a strategy for solution: this is the most difficult, and we shall come back to this particular question later,
- Applying the strategy: this is often simple.

4. *Validation*. This word "validation" is probably not used enough in the science and technology curriculum: the proof is that many students at the master level are not able to remember that they were told about it. There will be more about that below.

4.3 WHICH TOOL?

In the 1960s, the introduction of electronic calculators was considered a wonderful advance in comparison to the use of traditional pens (or chalk, when writing on blackboards), and even in comparison to the use of slide rules or logarithmic tables (Sandström, 2013). Later, the introduction of personal computers moved science toward the digital era. At first, the software was mainly used for numerical calculations, but the introduction of software for formal algebra (computer algebra systems) was a huge step forward.

Of course, many mathematicians still use blackboards, but some moved to using software for symbolic calculation, such as *Maple*, or *R*, or *Mathematica*, or *Maxima* (Chonacky and Winch, 2005; Steinhaus, 2008). This is not simply a question of writing properly, but also of efficiency. Certainly, using a text software is faster than using a pen, with cleaner results. Certainly, pocket calculators can give the result of a calculation; however, using software keeps track of the calculation, which is useful when validating it, or checking it when strange results are obtained.

This move toward computer algebra systems has huge consequences on the learning of science and of technology. For example, calculating a pH no longer calls for what was taught only a decade ago: now, you simply need to write down the equations (do the "chemistry") and use the "solve" command!

Students should not ignore the existence or the use of such software because they are so simple to use. In our research group, interns typically learn how to use *Maple* in less than one afternoon. Of course, they don't know all functionalities, but they can use it for simple writing and calculation. Later they find it very helpful for the following reasons:

- they simply use the mathematics that they learnt at school (the amount of new information is almost entirely restricted to understanding the difference between an egality "=" and an assignment ":=", something often learned already during computing courses),

- they become able to perform mathematical calculations that they may have learnt but could not perform easily (integration, differentials, etc.),

- in a single software, they get the possibility of typing text and making formal calculations as well as numerical ones, which is "real" scientific and technological life (the separation of the three activities is artificial, and a historical remnant of times when software was crude),

- calculating so easily gives them the possibility to explore calculations, to make more of them, to test ideas with more freedom. Although not all software are free, students have access to cheaper prices. (Compare the cost to the price of a pint of beer ;-)).

4.4 A FIRST SUMMARY

At this point, it is helpful to summarize the consequences already obtained in a "frame for calculation", encompassing all rules of good practice in calculation. The proposal is to copy and paste this "frame" each time you have to calculate, and then to follow the instructions one by one, from beginning to the end... until the final result is obtained, as if by a miracle.

But first, let us summarize the ideas that we proposed above:

1. it is preferable to use *Maple* (or another software of the same kind) rather than word processors or spreadsheets.

2. calculating means speaking (writing), making sentences in a natural language.

3. calculations have to be done in a structured way.

4. successive steps need to:

- express the problem in natural language,

- put down numerical data after translation in SI units,

- translate the problem into formal language through the introduction of the names of variables,

- solve the problem, which means that the problem in formal language is developed until the solution it is reached,

- introduce the numerical data in the formal solution,

- validate, discuss.

Accordingly, to aid students, they are invited to use the following document (Figure 4.2) (copying and pasting the model, then filling the slots). This will be done in this book.

4.5   THE QUALITY OF CALCULATION, VALIDATION

Finally, imagine that the solution is found. How do we know that a calculation is right? This question is important because the devil is hidden behind any calculation, even simple ones. Simple errors such as one forgotten "-" sign, a letter z confused with a 3, a bad operation (even when it is a simple one), switching two digits in a number, a bad proportionality...: all this can occur and does occur indeed.

It is funny to observe that remedies to such problems are seldom proposed in spite of their high frequency. Of course, in high school, one learns to check calculation by considering order of magnitudes, but this is a very crude validation. And although many professors tell students to make their calculations twice, the result is often that when a mistake is made once, it is made twice.

It is probably better to propose other solutions. For example, in physics, the units should be the same on both sides of equations, and this is the basis for the use of "dimensional analysis" (Gibbings, 2011). Symmetries, also, can be helpful. When the solution of an equation is found, injecting it into the equation should be common practice (as is proposed by *Maple*). For a formal equation, particular cases can be sought.

In summary:

- determine orders of magnitudes,

- make the calculation again entirely,

- apply the result of simple sets of data for which the results are known,

- use random data, and explore the solution space,

- test extreme cases,

- validate by another calculation.

Title of the calculation

Author
Date

▼ The question: in natural language, with the right words, in a concise and simple way (subject, verb, complement)

▼ Analysis of the question (often questions are solved immediately when they are analyzed)

　▼ The data is introduced: one has to write down the available data (for example digital values given). And immediately, the data are translated into units of the International System.

　▼ Qualitative model: make a scheme (not a beautiful picture) representing the phenomenon studied; then write down the assumptions that were used, including simplifications.

　▼ Quantitative model: describe formally the characteristics of the objects that you see on your picture, *i.e.*, introduce formal symbols (letters) that will be used for calculation.

▼ Solving

　▼ Looking for a solving strategy: how are you going to reach the solution (often this step is useless because the next one is enough)?

　▼ Implementing the strategy; if there is no strategy, one can look for relationship (equations) between the symbols, and often the solution is reached.

　▶ Discussion, validation (a second, or even a third calculation should be different from the first: one should not follow the same way as was initially done, because a mistake will be difficult to find)

▼ Expressing the results

　▼ The formal result found is written down again

　▼ Finding digital data (if possible with references; here, the initial data are copied)

　▼ Introduction of data in the formal solution

　▼ Discussion, playing with parameters in order to explore the solution space

▼ Conclusions and perspectives

　▼ Conclusions

　▼ Perspectives

FIGURE 4.2　*The framework for calculation.*

And now, equipped with all this background, we can embark on the study of the questions in the following chapters. They are organized by order of increasing physical complexity, as proposed by the "dispersed systems formalism", or DSF, as will be explained at the beginning of each chapter (This vo Kientza, 2021b).

Remember, it's all about training, with the right amount of enthusiasm and creativity... like going for a jog.

REFERENCES

Aguilera JM, Michel M, Mayor G. 2004. Fat migration in chocolate: diffusion or capillary flow in a particulate solid?—A hypothesis paper, *Journal of Food Science*, 69(7), R167–174.

Don't hesitate: read it and read it again.

Belitz HD, Grosch W, Schieberle P. 2004. *Food chemistry* (3rd ed), Springer Verlag, Heidelberg, Germany.

Bernardin L. 2022. *How Maple compares to Mathematica*, https://www.maplesoft. com/products/maple/compare/HowMapleComparestoMathematica.pdf, last access 2022-01-03.

For sure, Bernardin works for Maplesoft, but what he says is true. And you can find an answer by Mathematica online.

BIPM. 2020. *The International System of Units (SI)*, https://www.bipm.org/en/measurement-units/.

Yes, let's use SI units because we have to speak a common language!

Benson SW. 1965. Bond energies, *Journal of Chemical Education*, 42, 502–518.

Chonacky N, Winch D. 2005. *3Ms for instruction: reviews of Maple, Mathematica, and Mathlab, Computing in science and engineering*, https://www.researchgate.net/publication/232627402_3Ms_for_Instruction_Reviews_of_Maple_Mathematica_and_Matlab, last access 2022-01-04.

Ciqual. 2022. *French food composition table*, https://ciqual.anses.fr/, last access 2022-01-03.

About the information given in such tables, beware, because no two tomatoes are alike, for example.

Cottrell TL. 1958. *The strengths of chemical bonds*, 2nd ed, Butterworths, London, UK.

De Gennes PG, Brochard-Wyart F, Quéré D. 2003. *Capillarity and wetting phenomena: drops, bubbles, pearls, waves*, Springer, Heidelberg, Germany.

Duconseille A, Astruc T, Quintana N, Meersman F, Sante-Lhoutellier V. 2015. Gelatin structure and composition linked to hard capsule dissolution: a review, *Food Hydrocolloids*, 43, 360–373.

Fick A.1855. Ueber diffusion, *Annalen der Physik*, 94 (1), 59–86.

Gibbings JC. 2011. *Dimensional analysis*, Springer Verlag, London, UK.

Goutefongea R, Rampon V, Nicolas N, Dumont JP. 1995. *Meat color changes under high pressure treatment*, In: AMSA (Ed.), 41st ICoMST (II, 384–385), San Antonio, TX, USA.

IUPAC. 2019. *Colloid*, Compendium of Chemical Terminology, https://doi.org/10.1351/goldbook, last access 2022-01-03.

Mind that this is a comment, if change of character, there was a comment on SI units, as a common language. The same holds here in terms of words.

Jongsma C. 1979. *Axiomatic Structure and the Method of Analysis: Shifting Styles in the History of Mathematics*. A Second Conference on the Foundations of Mathematics, 3. Retrieved from https://digitalcollections.dordt.edu/faculty_work/305, last access 2022-01-03.

Joseph Fourier, *Théorie analytique de la chaleur*, Firmon Didot père et fils, Paris,1822.

Kopp J, Sale P, Bonnet Y. 1977. Contractomètre pour l'étude des propriétés physiques ds fibres conjonctives: tension isométrique, degré de réticulation, relaxation, *Canadian Institute for Food Science and Technology Journal*, 10(1), 69–72.

Lavoisier AL. 1790. *Elementary Treatise on Chemistry* (translation Robert Kerr), William Creech, Edinburgh, UK.

Remember that Lavoisier was the founder of the modern chemistry that we have today, as "chemystry" slowly evolved from alchemy, in the 17th and 18th centuries.

Lide DR (ed). 2005. *Handbook of chemistry and physics*, CRC Press, Boca Raton, FL, http://webdelprofesor.ula.ve/ciencias/isolda/libros/handbook.pdf, last access 2022-01-03.

Maplesoft. 2022. *Maple*, https://maplesoft.com/, last access 2022-01-03.

NCBI. 2022. *Tristearin*, https://pubchem.ncbi.nlm.nih.gov/compound/tristearin#section=Information-Sources, last access 2022-01-03.

PAC. 1996. *Density*, Glossary of terms in quantities and units in Clinical Chemistry (IUPAC-IFCC Recommendations 1996), 68, 968.

Piskounov N. 1996. *Differential and integral calculus, Mir, Moscow*, Soviet Union, https://archive.org/details/DifferentialAndIntegralCalculus_109, last access 2022-01-03.

It exists in any language. Select your own.

Reif F. 1967. *Statistical physics* (Berkeley Physics Course), 5. McGraw-Hill, New York, USA.

I love this one as well, even if it is not very new. As for the Feynmal course in physics, it is very clear and mind opening.

Sandström C. 2013. Facit and the displacement of mechanical calculators, *IEEE Annals of the history of computing*, CS-ANNALS-3503-130002.3d, 1–12.

Stefan J. 1879. *Über die Beziehung zwischen der Wärmerstrahlung un der Temperatur*, Sitzungsberichte der Kaiserlichen Akademie der Wissenschaften: Mathematisch-Naturwissenschaftliche Classe, 79: 391–428.

Steinhaus S. 2008. *Comparison of mathematical programs for data analysis*, http://www.scientificweb.de/ncrunch/, last access 2022-01-04.

This H. 2014. *Mon histoire de cuisine*, Belin, Paris, France.

This H. 2017. *Teaching document—DSR: frameworks guiding experimental work in science*, Notes Académiques de l'Académie d'agriculture de France, 8, 1–14, https://www.academie-agriculture.fr/publications/notes-academiques/n3af-2017-8a-teaching-document-dsr-frameworks-guiding-experimental, last access 2022-01-03.

This H. 2021. *La rigueur terminologique pour les concepts de la chimie: une base pour des choix de société rationnels*, Notes Académiques de l'Académie d'agriculture de France, 1, 1–15, https://www.academie-agriculture.fr/publications/notes-academiques/la-rigueur-terminologique-pour-les-concepts-de-la-chimie-une-base, last access 2022-01-03.

This vo Kientza H. 2021a. Osmosis in the kitchen, in Burke R, Kelly A, Lavelle C, This vo Kientza H (eds), *Handbook of Molecular Gastronomy,* CRC Press, Boca Raton, USA, 441–446.

This vo Kientza H. 2021b. Dispersed Systems Formalism (DSF), in Burke R, Kelly A, Lavelle C, This vo Kientza H (eds), *Handbook of molecular gastronomy*, CRC Press, Boca Raton, USA, 207-212.

This vo Kientza H. 2022. *Back of the enveloppe calculation*, CRC Press (to be published). -;).

Weisstein EW. *Sphere*, MathWorld—A Wolfram Web Resource. https://mathworld.wolfram.com/Sphere.html, last access 2021-12-31.

Wikipedia. 2022. *Zentralfriedhof Vienna-Boltzmann.JPG*, https://en.wikipedia.org/w/index.php?title=File:Zentralfriedhof_Vienna_-_Boltzmann.JPG&action=info, last access 2022-01-03.

CHAPTER 5

Gases (G)

IN THIS FIRST CHAPTER of problems, we see in the title the letter G after the word "gases". This calls for some explanation about the "disperse system formalism", or DSF: it is a wonderful way to describe physical-chemical systems. Symbols designate the phases of which they are composed and are linked by "operators", which describe the relationships between the phases.

In food, and also for formulation activities (making paintings, coats, cosmetics, etc.), there are classically four phases: gaseous phases (G), aqueous solutions (W, for "water"), fats in the liquid state (O, for "oil"), and solids (S). The operators are "/", for random dispersion of structures of one phase in another, "+" for a coexistence of more than one phase, "x" for intermixing of two phases, "σ" for superposition and @ for inclusion (This vo Kientza, 2021a). Of course, the DSF can also use other descriptions such as "dimensions", but these will not be used as these are not needed now in this book.

5.1 WHY IS IT GOOD TO KEEP MERINGUES IN CLOSED BOXES?

Before reading what follows, please consider that, starting from the questions given in the title of the sections, you are advised not to read further until you have spent a good deal of time looking for an answer. Not an answer "in the air", but always a quantitative answer, in terms of formal calculation or orders of magnitude, as mentioned earlier.

Do not hesitate to take time, to "ruminate", to carry the question with you for a period of time, in idle moments: when you walk down the street, when you are on the bus or the subway, when you are stuck in the presence of idiots, in too long meetings...

Of course, while looking for the solution, do not forget the structure proposed in Chapter 4!

5.1.1 An Experiment to Understand the Question

5.1.1.1 First Step: the General Idea

Meringues? I cannot imagine that you do not know that they are wonderful culinary preparations, mostly sweet, which are actually "solid foams", with, therefore, air

DOI: 10.1201/9781003298151-5

bubbles in a solid continuous network. They are classically made from egg whites that are whipped (the egg whites are turned into foam when the whisk introduces air bubbles in the liquid), then a large quantity of sugar is added while whipping, and finally cooking occurs, in order to make an external crust and also to dry the inside more or less, depending on personal taste (Bocuse, 2012).

They are very easy to do and helpful when you cook often because egg yolks are needed for custard, sauces, cakes, etc. What can we do with the remaining egg whites? In the past, cooks stored them in a basin at room temperature, in the kitchen, and they regularly organized sessions of meringues or macaroons (meringue with almond and stuffing) making. As said, the recipe of both preparations is easy because it calls simply for whipping the egg whites, adding powders (sugar, almond powder) and cooking until a crisp envelop forms around either a crisp or a soft inside. And later these pieces can be stuffed with a sweet preparation, such as jam. Meringues are easier to prepare than macaroons. Let us make some, using the following recipe:

1. heat the oven at 125°C;

2. whip egg whites until a firm foam is obtained;

3. add a spoon of sugar, and whip again;

4. repeat the addition of sugar and whipping until you reach a proportion of 60 g of sugar for 1 egg white;

5. when all the sugar is dissolved (no crystal can be observed, and the foam has a smooth and brilliant texture), put a portion on a cooking parchment, and cook at 125°C for 20 minutes;

6. reduce the temperature of the oven to 100°C and continue cooking for 1 hour, if a soft center is preferred, or 2 hours for crisp meringues.

   Tips for using meringues:

- in assembling meringues, form a vessel in which you put sliced strawberries, served with sugar and whipped cream, or Chantilly cream (whipped cream with sugar dissolved in the aqueous phase) (This vo Kientza, 2021b).

- break crisp meringues into fragments and add them to a fruit salad immediately before serving (not in advance, otherwise the meringue fragments loose their crispiness, as indicated in the cooking commandments).

5.1.1.2 *Questions of Methods*

Here, the core of the process is whipping egg whites. For the interpretation of the process, it is useful to know that egg whites are made of about 90% water, and 10% proteins (various kinds of proteins, of which ovalbumin is a major one, *i.e.*, representing 45% of the total protein content) (Belitz *et al.*, 2004). Proteins are the ingredients that make the difference between whipping pure water, and whipping egg whites. When you whip water, you can observe that the whisk pushes air into water, making bubbles. But when the motion of the whisk stops, the air bubbles move upward, and explode at the surface (This vo Kientza, 2021c). On the contrary, whipping egg whites introduces air bubbles that float, as well, but do not explode: obviously, the proteins make a resistant layer around the bubbles.

Now, it is easy to analyze that if you push air with one wire of a whisk, the number of bubbles that you create is twice less than pushing air with two wire loops, and ten times less than with ten wire loops. This explains why, when buying a whisk, it is good to count the number of threads: the more, the better (and if the whip handle is bigger, you will avoid cramps and be more efficient).

5.1.1.3 *Where the Beauty Lies*

For both processes of whipping water or egg whites, the color changes as more bubbles are introduced into the liquid: under daylight, a white color appears. However, the analysis is easier with egg whites because the bubbles are more stable.

Before whipping, egg whites are not white, but rather transparent and yellow, slightly greenish. You will be able to observe that bubbles that you create when you whip for only a few seconds are transparent (because air is transparent), and you can see the bubbles because you see the reflection of the light on their surface. If the light is white, the reflection is white. As more and more bubbles appear during whipping, more and more reflections make the whipped egg white appear whiter and whiter.

More generally, one can hold the idea that when a food preparation becomes white, it is often because air bubbles or oil droplets are introduced in a liquid phase (coagulation can also generate a white color, but for other reasons).

5.1.2 The Question

Imagine that one only knows that meringues are produced by whipping egg whites, adding sugar, and whipping again until no crystal appears (smooth consistency); next, the sweet foam is cooked in the oven, first at 125°C for about 20 minutes, in order to make a "crust", and then at 100°C, for one or two hours, so that the water from inside is evaporated (you see, I repeat the process to be sure of understanding it clearly).

After cooking, why is it convenient to store the meringues in closed boxes? About this question, a literal analysis is needed to prepare a calculation that makes the idea clearer. I invite you to look for yourself before considering the solution proposed below.

5.1.3 Analysis of the Question

5.1.3.1 *The Data is Introduced*

Here, we need some knowledge about egg whites. Let us use the following orders of magnitudes:

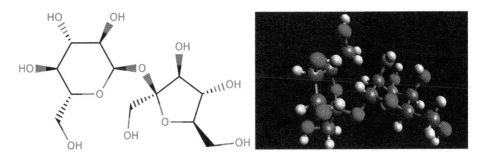

FIGURE 5.1 *Two representations of the molecule of sucrose (IUPAC: (2R,3R,4S, 5S,6R)-2-[(2S,3S,4S,5R)-3,4-dihydroxy-2,5-bis(hydroxymethyl)oxolan-2-yl]oxy-6-(hydroxymethyl)oxane-3,4,5-triol) (Juvin et al., 2019).*

- the mass of an egg is about 60 g;

- about half of the egg is egg white, *i.e.*, 30 g (some prefer to say 40 g, *i.e.*, 2/3 of the egg, but this does not make any difference when order of magnitudes are considered);

- the egg white is made of about 90% water (27 g $= 2.7\ 10^{-2}$ kg) and 10% proteins (3 g $= 3\ 10^{-3}$ kg).

For sugars, there are many, but white table sugar is almost pure (> 99.5) sucrose (Cultures Sucre, 2020), *i.e.*, a compound whose molecule is full of hydroxy groups (Figure 5.1).

Having given some general information, let us focus on the question studied: we assume that 50 g of sugar is added. Immediately, we have the reflex of converting this value to SI units. And even better, we do this step-by-step, even if we find it "obvious", because remember that being too fast is a frequent cause of errors: 50 g means 50×10^{-3} kg, or 5×10^{-2} kg.

When the egg white is whipped, the volume of foam is about 0.5 L: this means 5×10^{-1} L (look at the capital letter), or 5×10^{-1} dm^3, or 5×10^{-4} m^3.

Through whipping, air bubbles are introduced in the egg white: their diameter is visible at the beginning of the process (more than 1 mm), but it becomes so small that it becomes progressively invisible to the naked eye (This vo Kientza, 2021c): let us assume a final diameter of 0.1 mm (0.1×10^{-3} m, or 1×10^{-4} m).

Here, we have to add a comment about using software for formal calculation: in most popular programming languages (*Maple, Mathematica, Analytica*, C/C++, *FORTRAN, Matlab...*), an expression such as 6.02×10^{23} is equivalent to 6.02e23, or sometimes 6.02E23. Here, we shall use it frequently because it is a good training.

5.1.3.2 *Qualitative Model*

A qualitative model? We must learn to have a reflex: let us immediately construct a picture! More precisely, we have to construct two pictures: one with a meringue

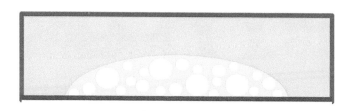

FIGURE 5.2 *A liquid foam is defined as a dispersion of gas structures (here, bubbles in white) in a liquid (blue). Some students ask how many bubbles are necessary for making a foam, and it can be answered that the system should be "colloïdal": "The term refers to a state of subdivision, implying that the molecules or polymolecular particles dispersed in a medium have at least in one direction a dimension roughly between 1 nm and 1 μm, or that in a system discontinuities are found at distances of that order" (IUPAC, 2018). At the bottom, we see the dried meringue in a box, with humid air above.*

in a closed box (Figure 5.2), and another one for a meringue outside a box. Between the two pictures, the difference is air: there is a small quantity of air in the box, and a large quantity in open air. And we know that air is a mixture of dinitrogen N_2, dioxygen O_2, carbon dioxide CO_2, water vapor H_2O, and other gases. Obviously, we are interested in water vapor, because—as we recall from the 6th cooking commandment—water without water is hard, but soft materials contain a liquid.

Let us now focus on the meringue itself. In Figure 5.2, the air bubbles are depicted as spherical, but simple optical microscopy of whipped egg whites shows that they can be polyhedral (Figure 5.3): *a priori* this should have little influence on softening.

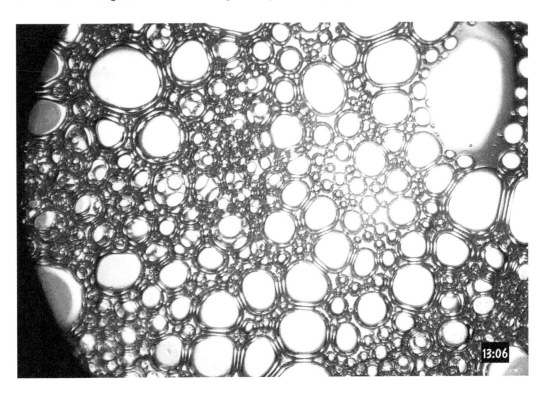

FIGURE 5.3 *When many air bubbles have been introduced into an egg white, they are so tightly packed that they lose their spherical shape and make polyhedrons.*

Now, let us move forward in the analysis of the initial question. After the cooking and the evaporation of water, the solid matrix of the meringues consists of proteins and sugars. Remember that a food product without water is crisp (one of the 14 commandments for cooking), whereas a food product full of water is soft. The meringues being made of proteins and sugar can attract water from their surroundings: remember the many hydroxy groups (-OH) in the formula of sucrose! They can make hydrogen bonds with water molecules.

Indeed, before cooking, these bonds were the reason why sucrose molecules could first dissolve in the aqueous phase of the foam. During cooking, a lot of energy (could you calculate how much, for example using the "latent heat of water vaporization" and the "arrow of energies" given in Chapter 4?) is needed to eliminate most water, while the "liquid foam" is transformed into a "solid foam". But when the meringues are stored later, humidity can soften meringues because all these hydroxy groups remaining are non-bonded.

If the meringues are stored in a box, they are in contact with some air, including some vapor, but if the quantity of air enclosed is small, then the quantity of water that can reintegrate the meringue is also limited.

Which quantity of water is present in the box? This is indeed the question being investigated here. Of course, this quantity has to be compared to the quantity needed to transform the sugar in the solid frame of the solid foam into a syrup.

For the dissolution of sugar in water, we know a limit: at room temperature 1960 g of sucrose can dissolve in 1,000 g of water. But of course, this would be the end of the story and not the beginning of softening. Looking at data online (Isengard, 2018), one finds that sugar contains about 0.2% water (2 g/kg): this makes a lower limit.

However, the exact quantity is not the question because when symbols are used instead of numerical data, in the end, one can decide about values and make various assumptions. About assumptions, I invite you to explicitly make some and discuss them.

Here, for simplicity, we make the arbitrary assumption that a meringue is badly damaged when a proportion p of water has been absorbed by the same mass of sugar. We can try to determine how much air is needed to deposit this mass of water in a meringue.

Finally, remember that this is only one possibility for calculation. Another solution could be based on the fact that the matrix of meringues is in a "glassy state", for which the state diagram can be used (Slade *et al.*, 1993).

5.1.3.3 Quantitative Model

For making the quantitative model, let us look at Figure 5.2 and simply characterize what we see.

For the whipped egg white, we assumed:

- 1 egg white, which means a mass M, including a mass M_w of water, and a mass M_p of proteins;

- the mass m of sugar.

For the air in the box, let:

- V be the volume of air in the box

- T be the absolute temperature

- P be the pressure.

Air can be assumed to be a mixture of dinitrogen, N_2, dioxygen O_2, carbon dioxide CO_2, water vapor H_2O, and other gases. We are interested in water vapor. It is known that, depending on days, there is more or less humidity: variations occur between 0 and the saturating vapor pressure P_s.

5.1.4 Solving

5.1.4.1 Looking for a Solving Strategy

Knowing the volume of the box and the vapor pressure, the mass of water in the box will be calculated.

Then, this mass will be compared to the mass of sugar.

5.1.4.2 Implementing the Strategy

In this section, we want to show that the solution is obtained almost automatically when one uses formal descriptions. Sometimes, the way toward the result is somehow incoherent but the most important point isn't that we reach the final result. Let us not be timid: let us try and try; do not hesitate to make something messy, because there is no pressure, and in the end, you will be able to "clean the desk", so that the final result is clean, free of the dross of its elaboration.

Indeed, we could visualize this work as if moving through a wild forest: we have to cut branches, turn to the left, turn to the right, until we cut through it. And later, when we see more clearly the route we have taken, we can find ways to go more directly from the starting point to the destination. To finish this discussion, let us say that this was the method of the great mathematician Carl Friedrich Gauss (1777–1855), who was referred to as the *Princeps mathematicorum*, the best mathematician: his published demonstrations were beautifully clean because he was not happy to deliver publicly his draft versions.

Let us now begin by calculating the mass of water in the air inside the box. We know that water in the air is there as vapor. And we know that vapor is a gas. It is only one of the various gases of air, along with dinitrogen, dioxygen, argon, carbon dioxide, and others. We learned that, in a gas made of several compounds, the total pressure is the sum of the partial pressures of the various constituents (McQuarrie and Simon, 1997).

We assume that all constituents behave as ideal gases:

$$P_i \cdot V = n_i \cdot R \cdot T$$

where i is the i-th gas in air by decreasing order of quantity (it is a good practice to rank objects non-arbitrarily). For water vapor, the value of i is often 3, depending on its humidity, but let us use an index w instead of a number (because the numbers i are useless) for this calculation:

$$P_w \cdot V = n_w \cdot R \cdot T$$

In this equation, we know V (the volume of the box), R (the constant of ideal gas), T (the absolute temperature of storage of the meringues), but how much is P_w? As mentioned, it can vary, depending on the place, time, etc. For simplicity, let us take as equal to the maximum, the saturation vapor pressure P_s:

$$P_s \cdot V = n_w \cdot R \cdot T$$

From this equation, one can easily determine the number of moles of water n_w:

$$n_w = \frac{P_s \cdot V}{R \cdot T}$$

Here, let us stop for a moment, in order to consider the new way of calculating using software. Certainly, at the university level and even earlier, we can ourselves divide both sides of the equation $P_s \cdot V = n_w \cdot R \cdot T$ by $R \cdot T$, and switch the left and the right side. But there is also the possibility to use the software to do it; below, with *Maple*, there was only need to click right on the equation and ask for isolation:

$$P_s \cdot V = n_w \cdot R \cdot T \xrightarrow{\text{isolate for n[w]}} n_w = \frac{P_s V}{RT}$$

Why do so? Because it is validation! The answer is not whether to use one method instead of another, but to use the two, in order to be surer of ourselves.

Generally, in this book, you will see this philosophy in practice. The blue formulas are the results given by the software.

That said, let us move one step forward: if we know the number of moles of water n_w, we can determine the mass of water in the air m_w by the equation:

$$m_w = n_w \cdot Mm_w$$

where Mm_w is the molar mass of water $(16 + 1 + 1 = 18 \text{ g})$.

So that the mass of water in the air:

$$m_w = \frac{P_s V}{RT} Mm_w$$

5.1.4.3 Validation

Here, we shall observe that validation is a difficult question because it sometimes takes much creativity to find a different calculation; at times, there is only one way to reach the result.

For example, without searching for too long, we can:

- do the same calculation on a different medium (paper, blackboard...) because changing the medium sometimes gives a different perspective;

- do the same calculation numerically, using orders of magnitudes;

- consider units, in the result (BIPM, 2022):

$$m_w = \frac{P_s \cdot V}{RT} \cdot Mm_w$$

Let us use now the last possibility.

A pressure is in pascals (Pa, or J/m^3); a volume is in m^3; a molar mass in $g.mol^{-1}$; R is in $J \cdot mol^{-1} \cdot K^{-1}$; and T is in K.

For the term at the right of the equal sign, the units are:

$$\frac{J \cdot m^{-3} \cdot m^3 \cdot g \cdot mol^{-1}}{J \cdot mol^{-1} \cdot K^{-1} \cdot K}$$

This simplifies in g, which is indeed the units of the mass m_w (by the way, remember that the gram, g, is not a basic unit of the SI).

5.1.5 Expressing the Results

5.1.5.1 The Formal Result Found is Written Down Again

This step seems obvious and it is: we simply repeat the previous result.

$$m_w = \frac{P_s \cdot V}{RT} \cdot Mm_w$$

Why repeat it? Because we need to have the formula under our eyes directly, to enjoy it. It gives us time to think about what we can do with it: it will be compared to either the total mass m of sugar in the meringue, or the mass of sugar in the crust if any (a proportion of m).

5.1.5.2 Finding Digital Data

Let us look at the solution: we have to find the saturating vapor pressure in scientific sources, such as a handbook of physical and chemical constants, or, today, on the Internet.

We find that, at 20°C, it is 2.3 kPa (Hardy, 1998). But before moving on, it can be helpful for one's instruction to compare this value to the total pressure of the atmospheric air: about 10^5 Pa, or 100 kPa: the water vapor contributes up to 1/40 of the pressure of air.

Now we can write, using our previous formalism:

$$P_w = 2.3\text{e}3 \text{ Pa}$$

Let us assume that the dimensions of the box are 50 cm × 50 cm × 10 cm. The volume would be (with SI units, always with SI units):

$$V = 0.5 \cdot 0.5 \cdot 0.1 = V = 0.025 \text{ m}^3.$$

Remember: we see a blue color in the result, because the product was evaluated using the evaluation command of the software. By the way, let us observe that, without special care, this is evaluated with an uncontrolled number of digits. For example, if we write $\frac{4.0}{7.0} = 0.5714285714$, the number of digits implicitly fixed (here 10) is displayed. On the other hand, one can choose an option for displaying the result in scientific notation, such as in

$$\frac{4.0}{7.0} = 5.71 \times 10^{-1}$$

Coming back to our calculation about meringues, we know that $R = 8.31$ J · mol^{-1}· K^{-1}(Hardy, 1998), Mm_w: 18 g = 18 × 10^{-3} kg.

5.1.5.3 Introduction of Data in the Formal Solution

With software, there is a command (*subs* in *Maple*) for introducing numerical data in an expression. Let us use it:

$$subs\left(V = 0.025, R = 8.32, P_s = 2.3\text{e}3, T = 293, Mm_w = 18e - 3, m_w = \frac{P_s \cdot V}{R \cdot T} Mm_w\right)$$

$$m_w = 4.25 \times 10^{-4}$$

Because we used here the value of the molar mass in kg · mol^{-1} (instead of the usual g.mol^{-1}), this result is in SI units, *i.e.*, kg: the maximum quantity of water in the box is about half a g, and this is to be compared to 50 g of sucrose (about 1%).

5.1.5.4 *Discussion: Playing with Parameters in Order to Explore the Solution Space*

Boxes with the dimensions used in the above calculation are frequent in kitchens. What about leaving meringues in the open air of the room, which amounts to a box of volume 5 m × 5m × 3 m (for example)? Then the mass of water (in kg) would be:

$$subs\left(V = 75, R = 8.32, P_s = 2.3e3, T = 293, Mm_w = 18e - 3, \frac{P_s \cdot V}{RT} \cdot Mm_w\right)$$
$$1.27 \times 10^0$$

More than one kilogram: this is certainly a bad result for meringues because it is more than enough to make a saturated syrup.

5.1.6 Conclusions and Perspectives

In the above calculation, we did not use the phase diagram showing the glass transition of sucrose solutions (Champion *et al.*, 1997), but the question about meringues provided the opportunity to observe how simple ideas about gas behavior can be useful, as a first approach to a question. Also, it was useful to see how results from the scientific curriculum can be applied in daily (culinary) life.

One observation about scientific practice: in the past, chemistry laboratories were equipped with a thermometer and a hygrometer, to record temperature and humidity at the beginning of the day. These tools are seldom present today, so that young engineers or scientists do not perceive much humidity, and we are less conscious of their variations than before. Therefore, we sometimes forget to take it into account, which can be a handicap when we consider the consistency of food, whose variations are so important: the human brain detects constrasts, of colors, of tastes, of consistencies, etc.

5.2   WHY IS IT HOT OVER STOVES?

5.2.1   An Experiment to Understand the Question

5.2.1.1 *First, the General Idea*

The question given in the title of this section is obvious: the hot stove heats the air around, and above, in particular. But remember that this is not the kind of answer that we are looking for: instead we want to invent a small calculation concerning the question. Moreover, the answer given above deserves more analysis: what is "hot"? why does the stove heat the air around? why does it heat the air above "in particular"?

For the experiment about the question of this chapter, a simple stove is needed: place a light sewing thread at some distance over the hot surface, and you will see that (depending on the power of the stove) it moves erratically or is lifted up.

Having shown that hot air goes up, try to look at an object behind the stove, looking just over the surface, along a line parallel to the surface of the stove: you will see that the object appears blurred.

5.2.1.2   Questions of Methods

In order to explore this phenomenon but in liquid phase, prepare some coffee, putting ground coffee in a filter, and adding more and more water until the percolating liquid is clear, with no color, whereas the powder remains black.

You can then put this dark powder in a glass that you will heat on the stove, for example by tilting the glass so that only one small part is heated at the bottom.

Using this experimental set up, the dark powder will help you to see the heated liquid moving up, and back down along the colder walls of the glass. This motion corresponds to "convection".

5.2.1.3   Where the Beauty Lies

It is useful to know the result of the previous experiments when you use a scale, in the laboratory, because it explains why textbooks advise not to weigh hot objects.

Let us put their advice to the contrary: put a piece of hot cooked egg white on a precise scale, and you will see that the displayed mass is decreasing. Certainly cooked egg white contains 90% water (see section 5.1), and hot water can evaporate at room temperature, but if you do the measurement with a hot metallic spoon, the same kind of displayed mass decrease is observed! Obviously, the metal is not "evaporating" at a temperature less than 100°C (do you have an idea of the vapor pressure of iron at 20°C, for example?), and another phenomenon is certainly occurring.

The previous experiments provide the key to the mystery: air heated by a hot object rises; but, as the gas is accelerated, its pressure is reduced (this is the "Bernouilli model") (Feynman *et al.*, 1963), so that the plate of the scale is moved from high to low pressure (remember that a fluid has a zero velocity near a surface). As a consequence, this lift apparently reduces the weight of the object on the scale.

Now, we know for sure that we should avoid weighing hot objects, and we can understand why it is so important to apply rules such as the ones that you will find in the book of official methods of AOAC, the Association of Official Analytical Chemists (AOAC, 2019). In this book, it is rightly said that when you want to weigh a hot object, you can either put it in a closed cold box or first store the hot object in a dessicator, so that it cools without its mass being modified by humidity from air.

5.2.2   The Question

It is a good advice to give to students to repeat a question with one's own words: it air hot over stoves; why is the stove hot, and why is the air over the stove also hot?

5.2.3   Analysis of the Question

5.2.3.1   The Data is Introduced

No data needed. So why keep this section? Because we need training: repeat until the steps of the process are a refex.

5.2.3.2 Qualitative Model

First, let us digress about questions and, in particular, culinary ones. Too often, culinary books are full of instructions that were followed by cooks for decades or even centuries without being investigated. For example, in the 17th century, it was written that a steel bar under a barrel could prevent the wine in the barrel from "turning" during a storm (Bonnefons, 1655); for years it was said that women on their periods would ruin the mayonnaise (This, 2010); for decades, until recently, it was said that the volume of whipped egg white is larger when whipping was performed always in the same direction. And so on: for French cuisine, I collected tens of thousands of these "culinary precisions", a term that covers sayings, proverbs, tips, tricks, old wive tales and so on.

Rational minds were able to perceive that many of these culinary precisions were wrong, but what is better way than an experimental test? One counter example is indeed enough to refute a general law. Certainly, let us conduct experiments before trying to explain phenomena; and only when the phenomena are clearly established, let us look for explanations.

About the stove, it is important to repeat that the "heat" corresponds to molecular and atomic motion, but also to infrared radiations that propagate in every direction, and not only upward as is too often assumed, in particular when people make barbecues and place the meat over burning charcoal.

A proof that infrared radiation goes in any direction? Put one hand close to your cheek, and you will perceive heat by the cheek, be your hand and cheek vertical or horizontal (when you tilt your head). And obviously you also know that it is hot in front of a fire, and not only over it. Indeed, placing meat that you wish to cook on the side of a fire, and not over it, is as fast (the only question is distance, because the energy received by a surface decreases as the square of the distance to the hot source), and it is safer, because you avoid toxic benzopyrenes that the burning wood or charcoal would deposit on the meat (Park *et al.*, 2017). Moreover, putting the meat on the side of the fire, rather than above it, has the other advantage that you can put a drip pan under the meat, allowing you to recover the wonderful juices that are released (meat shrinks when cooked over 55°C)(Lepetit, 2008) on bread or potatoes that you place in the pan.

This being said, let us come back to the question, assuming a stove has a flat hot surface on the upper part, and air over it (Figure 5.4).

Initially, the air is at room temperature, but we know that the air just over the hot surface is heated, in part by radiation: remember the Stefan equation. The infrared radiation emitted by the hot surface of the stove "hits" molecules in the air (perhaps it would be more accurate to say "there is a radiation/matter interaction), so that they get energy, and the temperature of the air increases. And we all "know" that hot air goes up, but why? After all, hot air *could* go down.

The group of exercises that we shall consider now is indeed a demonstration of simple physical equations, but it is good to remind students that they can often use their own resources to fix their ideas and find formulas that they forgot.

Here, the starting point is to observe that the stove is heating air around so that it produces a volume of hot air in the colder air of the room. We also know that

FIGURE 5.4 *Air over a flat hot surface. Such a picture seems useless, but making it is training for situations where diagrams will be necessary. Moreover, the very fact of representing the various objects that one considers is an invitation to characterize them quantitatively.*

matter expands generally with increasing temperature (if needed, one could make the complementary experiment of filling in a precision burette in a cold room and transferring it in a hotter place.

Even if you do not see how to solve the initial question, it is good that you observe that this solution will come quite automatically when the question "how much" is asked.

5.2.3.3 Quantitative Model

Here there is no need to define the mass of air, and we have to consider different temperatures: the ambient temperature T_{amb} and the temperature of hot air just over the hot surface T.

5.2.4 Solving

5.2.4.1 Looking for a Solving Strategy

We guess that hot air expands, but why? We shall examine this question.

Then because hot air expands, the distance between two neighboring molecules increases. But how much?

And finally, hot air goes up: this involves buoyancy. We shall find the force lifting hot air.

5.2.4.2 Implementing the Strategy

If we look at the 14 definitions and formulas that were given in the chapter 4, we see that some could apply because they have a relationship with heat:

- the definition of density (for air).

- the definition of kinetic energy (because air is moving).

- the model for ideal gases (because air is gaseous).

- the Stefan model (because the stove radiates).

- the Fourier model (because there are heat transfers).

However, we can probably eliminate the Stefan model because air is not a black body, and the Fourier model, because the question is not of finding how heat is transferred. About kinetic energy, this would characterize the motion, but it would not explain why the air moves. So, we are left with the definition of the density and the formula of ideal gases.

The latter is written as follows:

$$PV = nRT$$

And it can be used to determine density, when the temperature is increased. For example, for a certain volume V, we would have a certain number of moles:

$$PV = nRT \xrightarrow{\text{isolate for n}} n = \frac{PV}{RT}$$

Also the density is:

$$\varrho = \frac{m}{V}$$

If we have the number of moles, on one hand, and the mass on the other, we understand we can link them using the Avogadro number N_A (this was not one of the 14 definitions and equations, because it is too basic, for chemists!):

$$m = N_A \cdot n = N_A \frac{PV}{RT}$$

So that the density is:

$$\varrho = \frac{m}{V} = N_A \frac{P}{RT}$$

We see here what we would have guessed: when the temperature T increases, the ratio decreases: density is decreased.

But a fluid of lower density in another fluid is moving toward the direction opposite to its weight, because "buoyancy" becoming greater than the weight.

Why buoyancy? See Annex 3.

5.2.5 Annexes

5.2.5.1 Annex 1: How far apart are molecules in air?

In the previous calculations, it is calculated that heated air can expand, so that the distance between two neighboring molecules is increased. But how far apart are air molecules?

5.2.5.1.1 The Question

Let us simply repeat: how far apart are molecules in air?

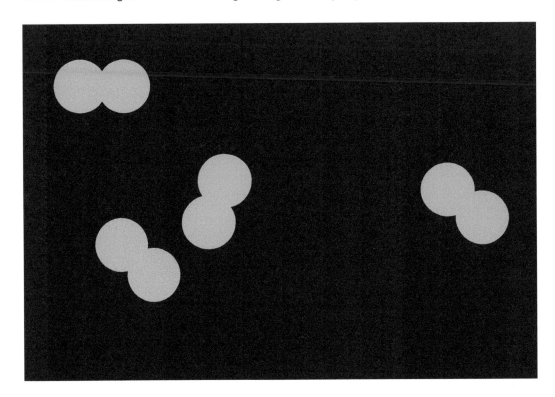

FIGURE 5.5 *Air at a molecular level. In such representations, diatomic molecules are shown, and of course for air the majority of them are nitrogen N_2. Because there is nothing (in particular no "air") between the molecules of air, the background has to be black. A main issue is the average distance between molecules. In this* a priori *picture, the distances (compared to the radius of molecules) are arbitrary.*

5.2.5.1.2 Analysis of the Question

5.2.5.1.2.1 The Data Is Introduced Here again, no data is given, and we have to find everything by ourselves.

5.2.5.1.2.2 Qualitative Model We have to represent air: Air is a complex mixture, but it is a good strategy to look for simplification, considering that it is only dinitrogen N_2 (because air is mainly that).

In Figure 5.5, we have to show N_2 molecules quite far apart in the vacuum. But how far apart? Because we do not know, our first representation is arbitrary; however, we shall create an improved picture after the question is answered.

In this picture (a snapshot, because we know that molecules in a gas are at random distances, and moving in all directions), the background has to be black, because there is nothing there. The size used for the representation of molecules is arbitrary, and their axis are in random directions; the distances between pairs of molecules are randomly distributed as well.

FIGURE 5.6 *Scheme for a "lattice calculation" of the distance between molecules in a gas. Here again, because the picture is drawn before the calculation, the distances are unknown. This calls for another picture, after calculating.*

However, it is good practice to make a simple description first: we shall simplify as much as possible, in terms of dimensions of space, of shape of objects, of their distribution, etc. Here all molecules will be dots, and they will be organized in a cubic grid (Figure 5.6).

5.2.5.1.2.3 Quantitative Model As previously, we simply need to introduce formal parameters that will describe the picture.

In Figure 5.6, we see a volume V in which there is a number N of molecules, each having a radius r. The distance between two molecules will be written d. And because we decided to consider dinitrogen molecules, we characterize the chemical nature by the molar mass M.

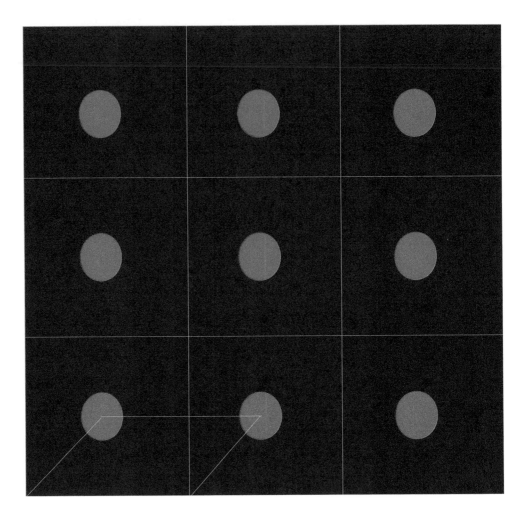

FIGURE 5.7 *Each molecule occupies a certain elementary volume. The distance be-tween the centers of two neighboring molecules is equal to the size of elementary volumes (cubes, for a volume).*

5.2.5.1.3   Solving

5.2.5.1.3.1   Looking for a Solving Strategy   We can divide the space into elementary cells, each containing one molecule (Figure 5.7).

From this picture, it should be clear that the volume of one cell is equal to the total volume V of the gas divided by the number N of molecules.

Knowing the volume of elementary cubes, it is easy to find their size (the side of the cubes). This means that we shall be able to determine the distance between the center of two neighboring molecules.

5.2.5.1.3.2   Implementing the Strategy   From the previous analysis, we see that we need first to know the number of molecules in a certain volume. And again, because we consider a gas, the scanning of the list of definitions and formulas shows that the formula for ideal gases can again be helpful:

$$PV = nRT$$

Here, we can decide for a volume. We know the pressure (atmospheric one), the temperature (room temperature), the constant R, and we see that the number of moles n in the volume V can be determined by:

$$n = \frac{PV}{RT}$$

Remember that what we are looking for is the number N of molecules. Of course, the number of molecules is given from the number of moles using the Avogadro number:

$$N = nN_A$$

So that:

$$N = \frac{PV}{RT}N_A$$

And we remember that the volume v associated with one molecule (a small cubic box in the center of which is one molecule) is equal to the volume V divided by the number N of molecules:

$$v = \frac{V}{N} = \frac{V}{\frac{PVN_A}{RT}} = \frac{RT}{PN_A}$$

Finally the distance between two neighboring molecules is equal to the side of this volume, that we find with:

$$d = \sqrt[3]{\frac{RT}{PN_A}}$$

5.2.5.1.4 Expressing the Results

5.2.5.1.4.1 The Formal Result Found Is Written Down Again Some students ask why one should write again the final result, and we already explained one reason. There is another one: because calculation is like a language, a "conclusion" is always helpful. It makes it clearer for others–and for oneself–to see that the goal was reached.

And here it is, accordingly:

$$d = \left(\frac{RT}{PN_A}\right)^{1/3}$$

5.2.5.1.4.2 Finding Digital Data (If Possible with References; Here, the Initial Data Are Copied) We know the value of the constant $R = 8.32$ J · mol^{-1} · K^{-1}. We assume that $T = 300$ K. The pressure is chosen to be about $P = 10^5$ Pa (for sure, more significant figures could be used, but because of the assumptions that we made, they are useless).

The Avogadro number is chosen to be 6×10^{23}.

5.2.5.1.4.3 Introduction of Data in the Formal Solution Below, the *evalf* command of the *Maple* software determines the numerical value of the expression in brackets, and the *subs* command gives the value of the last formula, using the data given before it):

$$evalf\left(subs\left(R = 8.32, T = 300, P = 1e5, N_A = 6e23, \left(\frac{RT}{PN_A}\right)^{1/3}\right)\right)$$

$$3.4635676640 \times 10^{-9}$$

Here, because we chose SI units, the result is in m.

The result given by *Maple* (or other formal languages) can have as many digits as wanted. But it is good practice to avoid excessive precision and follow the scientific rules, which means giving only the significant digits.

However, because it is a fact that many students find this hard, one possibility (only for a first approach) can be to use a "scientific notation":

$$3.46 \times 10^{-9}$$

Now, let us take some time to investigate the better practice of mastering the calculus of the number of digits. It is important to know that for a formal expression $f(x_1, x_2, ..., x_n)$ (a function of n variables), the International Organization for Weights and Measurements decided that the uncertainty on the values of the function can be expressed as:

$$\Delta f = \sqrt{\sum_{i=1}^{n} \left(\frac{\partial}{\partial x_i}(x_1, x_2, ..., x_n)^2 \cdot \Delta x_i^2\right)}$$

the Δx_i being the uncertainty of each variable x_i (BIPM 2012).

I know that what is written above is indigestible: first, it is a complex general case, and secondly there are partial derivatives! Let us explain it simply, considering a solution made of a solute, such as sucrose, in water as the solvent. The concentration in sucrose, in this simple solution, can be described by the ratio:

$$c = \frac{m}{M}$$

where m is the mass of sucrose and M the mass of solute.

Here, c is a function of the two variables m and M:

$$c(m, M) = \frac{m}{M}$$

In this simplecase, the official formula is:

$$\Delta c(m, M) = \sqrt{\sum_{i=1}^{2} \left(\left(\frac{\partial}{\partial x_i}c(m, M)\right)^2 \cdot \Delta x_i^2\right)}$$

with $x_1 = m, x_2 = M$.

If you are weak in maths, the software can calculate the partial derivatives for you:

$$\frac{\partial}{\partial m}(c(m, M)) = \frac{\partial}{\partial m}\left(\frac{m}{M}\right) = \frac{\partial}{\partial m}c(m, M) = \frac{1}{M}$$

And:

$$\frac{\partial}{\partial M}\left(\frac{m}{M}\right) = -\frac{m}{M^2}$$

Let us introduce these values in the square root:

$$\Delta c(m, M) = \sqrt{\frac{1}{M^2} \cdot \Delta m^2 + \left(-\frac{m}{M^2}\right)^2 \cdot \Delta M^2}$$

You see: it is very simple.

Now, coming back to our question about the digits in the expression of the distance between dinitrogen molecules, we can consider the difference between 8.32 and the more precise value 8.314 462 618 153 24 for R, the difference between 10^5 and 103325 Pa for the atmospheric pressure, and the difference between the value 6e23 of the Avogadro number, and a more precise value.

We would write:

$$\Delta d = \left(\left(\frac{\partial}{\partial R}\left(\left(\frac{RT}{PN_A}\right)^{1/3}\right)\right)^2 \cdot \Delta R^2 + \left(\frac{\partial}{\partial T}\left(\left(\frac{RT}{PN_A}\right)^{1/3}\right)\right)^2 \cdot \Delta T^2\right.$$
$$+ \left(\frac{\partial}{\partial P}\left(\left(\frac{RT}{PN_A}\right)^{1/3}\right)\right)^2 \cdot \Delta P\right)$$
$$\left.+ \left(\frac{\partial}{\partial N_A}\left(\left(\frac{RT}{PN_A}\right)^{1/3}\right)\right)^2 \cdot \Delta N_A^2\right)^{1/2}$$

Using the values given:

$$evalf\left(subs\left(\Delta R = 0.01, \Delta T = 1, \Delta P = 1e4, \Delta N_A = 0.1e23, R = 8.32,\right.\right.$$

$$N_A = 6e23, T = 300, P = 1e5, \Delta d$$
$$= \left(\left(\frac{\partial}{\partial R}\left(\left(\frac{RT}{PN_A}\right)^{1/3}\right)\right)^2 \cdot \Delta R^2 + \left(\frac{\partial}{\partial T}\left(\left(\frac{RT}{PN_A}\right)^{1/3}\right)\right)^2 \cdot \Delta T^2\right.$$
$$+ \left(\frac{\partial}{\partial P}\left(\left(\frac{RT}{PN_A}\right)^{1/3}\right)\right)^2 \cdot \Delta P\right)$$
$$\left.\left.\left.+ \left(\frac{\partial}{\partial N_A}\left(\left(\frac{RT}{PN_A}\right)^{1/3}\right)\right)^2 \cdot \Delta N_A^2\right)^{1/2}\right)\right)$$

$$\Delta d = 2.20 \times 10^{-11}$$

Here, we determine an uncertainty of about 1×10^{-10}, for a value 3.46×10^{-9}: it would be reasonable to say that on the average, two molecules in air are separated by about 3.5 nm.

5.2.5.1.4.4 Validation Let us observe that one could also calculate the same result starting from the idea that one mole of gaseous matter has a volume of 22.4 L at 0°C (or preferably 24 L at 20°C). So that the volume per molecule is: $\frac{22.4e-3}{6e23} \approx$ 3.73 10^{-26}. The cubic root would give the same result as before.

5.2.5.1.5 Conclusions and Perspectives

In the previous calculation, some parameters that we introduced were not used: the molar mass was useless. This is in accordance with the assumption that any gas occupies the same volume in the same conditions (when it is in the room conditions).

Now, we observe that the distance between two molecules is about 3 nanometers, and this is to be compared with the size of diatomic molecules: it is good to remember that the simplest covalent bond (carbon-carbon) is about 0.15 nanometers long (McQuarrie and Simon, 1997). This means that, at room temperature and pressure, the distance between two neighboring molecules in air is 20 times their diameter: this allows to redraw the initial figure, with the right size.

5.2.5.2 *Annex 2: Why Gas Dilatation with Temperature?*

5.2.5.2.1 The Question

Heated air expands, we said, but why? Here this question is a way to help students to (re) investigate simple notions of statistical physics.

It is probably good to give first a frame for solving the general question. Let us observe that indeed the question was solved in the previous annex, because the distance between molecules was expressed as a function of temperature, and it was shown to increase with it.

However, this was not an explanation, but only a result in terms of calculation, and here we prefer to go down to deeper concepts, such as molecular motion, instead of state parameters.

5.2.5.2.2 Analysis of the Question

5.2.5.2.2.1 The Data Is Introduced No data is given.

5.2.5.2.2.2 Qualitative Model Let us consider a certain volume of air. In order to isolate it, we put it in a box. However, if the dilatation is to be considered, this box should be able to increase in size... and this is why courses in thermodynamics envision gases in cylinders closed by a moving wall.

This can be depicted as follows (Figure 5.8):

The question of calculating the pressure as a function of the temperature is a very classic one. Here the originality lies in organizing the calculation with a series of questions.

5.2.5.2.2.3 Quantitative Model In classical thermodynamics, one would introduce:

- *n*: the number of moles of gas in the cylinder

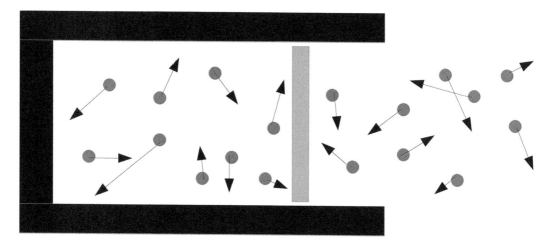

FIGURE 5.8 *Gas in a cylinder for calculating the pressure.*

- V: the volume of the gas

- P: its pressure

- T: the absolute temperature

What about our initial definitions and formulas? Of course, about a question such as the one considered here, the ideal gas model would apply, as said, but if we want to consider the motion of molecules, then the expression of the kinetic energy is a better choice.

In the volume, we have N molecules having each an average kinetic energy:

$$e_c = \frac{1}{2}mv^2,$$

where m is the mass of a molecule, and v the mean velocity (yes: average, because some are moving faster than others).

5.2.5.2.3 Solving

For this annex, we propose a slightly different game: you are invited to answer a series of simple questions, so uthat in the end you will reach the goal. Let us go (do not look immediately to the answer).

1. If the air inside the cylinder is at the same temperature as the air outside, why does the moving wall stand still?

   A: The wall is in equilibrium when the number of molecules bumping against it from the inside of the cylinder is equal to the number of molecules bumping against it from the outside. Collisions correspond to "pressure".

2. What happens when the gas inside the cylinder is heated (imagine a heating resistance in its wall)?

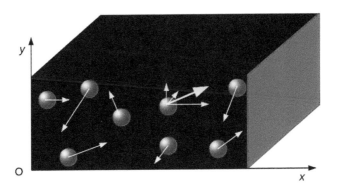

FIGURE 5.9 *In the box with a moving wall (red), molecules move in all directions. For each molecule, the velocity v has components along Ox, Oy and Oz.*

A: Heating the gas inside the cylinder means giving energy to the inner molecules, so that they repel more the moving wall than the molecules from the outside, and the volume of gas in the cylinder expands.

3. How much?

   A: Imagine a closed cubic box of volume V, with N molecules inside. One corner of the box defines a referential (O, x, y, z). Let us consider the wall of the box perpendicular to the Ox axis (Figure 5.9).

4. Consider the collision of one molecule against the moving wall. How would its velocity change?

   A: During the collision, only the x component of the velocity is inverted, from v_x to $-v_x$.

5. And how would the momentum change?

   A: The change in momentum is:

$$\Delta p = mv_x - (-mv_x) = 2mv_x$$

6. Can you now calculate the force exerted by the colliding molecule?

   A: The force exerted by the molecule is:

$$f = \frac{\Delta p}{\Delta t} = \frac{2mv_x}{\Delta t}$$

7. Now, we need to find Δt. We consider that there is no force when the molecule does not hit the wall. Can you calculate an average force, considering Δt as an average time between collisions with the moving wall?

   A: Δt is the time it would take the molecule to go across the box and back (a distance 2 l, if l is the length of the box in the Ox direction) at a speed of v_x.

Thus:

$$\Delta t = \frac{2l}{v_x}$$

And the expression for the force becomes:

$$f = \frac{2mv_x}{\left(\frac{2l}{v_x}\right)} = \frac{mv_x^2}{l}$$

8. This force is for one molecule only. What is it for N molecules?

   A: If we multiply by the number of molecules N and use their average squared velocity, we find the force:

$$F = N\frac{\overline{mv_x^2}}{l}$$

The bar means the average value.

9. Can you now find the expression in function of the velocity v rather and in function of v_x?

   A: we know that:

$$v^2 = v_x^2 + v_y^2 + v_z^2$$

The average squared velocity is:

$$\overline{v^2} = \frac{1}{N} \cdot \left(v1_x^2 + v1_y^2 + v1_z^2 + v2_x^2 + v2_y^2 + v2_z^2 + \cdots + vN_x^2 + vN_y^2 + vN_z^2\right)$$

Here $v1$ is the velocity of the first molecule, $v2$ the velocity of the second one, etc., vN being the velocity of the N-th molecule.

We arrange the sum as:

$$\overline{v^2} = \frac{1}{N}\cdot\left(v1_x^2 + v2_x^2 + \cdots + vN_x^2\right) + \frac{1}{N}\cdot\left(v1_y^2 + v2_y^2 + \cdots + vN_y^2\right) + \frac{1}{N}\cdot\left(v1_z^2 + v2_z^2 + \cdots + vN_z^2\right)$$

$$\overline{v^2} = \overline{v_x^2} + \overline{v_y^2} + \overline{v_z^2}$$

Because the velocities are random, their average components in all directions are the same:

$$\overline{v_x^2} = \overline{v_y^2} = \overline{v_z^2}$$

Thus:

$$\overline{v^2} = 3\overline{v_x}^2$$

or

$$\overline{v_x}^2 = \frac{1}{3}\overline{v^2}$$

10. Introduce this in the expression for F.

    A: The total force on the moving wall is the force for one molecule multiplied by the number of molecules

    $$F = N\frac{m\overline{v^2}}{3l}$$

11. And find the pressure.

    A: We simply use the definition of the pressure:

    $$P = \frac{F}{A} = N\frac{m\overline{v^2}}{3lA} = \frac{1}{3}\frac{Nm\overline{v^2}}{V}$$

12. You see that this expression looks like the ideal gas equation. If the kinetic energy of one molecule is equal to $\frac{3}{2}k_BT$, determine how this pressure is linked to temperature (Starzak, 2010).

    A: Here we see a piece of information that was not in the group of 14 definitions and formulas. But remember that we also announced that other results would be introduced when needed. So that we can write:

    $$e_c = \frac{3}{2}k_BT = \frac{mv^2}{2}$$

    for ideal gases, showing that volume increases with temperature.

5.2.5.2.4 Conclusions and Perspectives

Here from a series of simple questions, the students find the way for expressing the pressure of a gas as a function of the temperature. And it is clear that if the pressure increases inside the cylinder, the moving wall will indeed move, toward the side with less pressure, *i.e.*, the volume of the heated gas inside will increase.

5.2.5.3 *Annex 3: Buoyancy*

Some students do not remember the expression of buoyancy, and it is probably good to refresh their ideas, asking questions such as why is there some creaming of mayonnaise (emulsion of oil in an aqueous phase made of egg yolk and vinegar), or draining of water in a whipped egg white... and upward motion of hot air.

First let us observe that he same force acts on any fluid, liquid or gas. And we can be very general by considering structures with complex shapes.

Then it is a fact that whereas some students remember the expression of buoyancy, not all remember how this expression is obtained. Again, moving by elementary steps is helpful.

1. Make a sketch in order to represent a complex object in a fluid (Figure 5.10).

2. Which forces apply to the immersed object?

    A: The forces on the object are the weight P and the pressure forces exerted by the liquid on the surface (remember water pressing on your ears when you dive).

FIGURE 5.10 *A solid in a liquid. The first picture for this problem is not very interesting, except that it is an invitation to applying a strategy.*

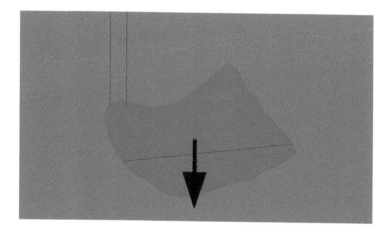

FIGURE 5.11 *The weight (black arrow) and the pressure forces (not shown). When the question is about forces, simple knowledge is to be applied.*

3. Why do these pressure forces act on the surface? Consider a small element of surface at a depth h. Calculate the force on it.

   A: Let us consider a small surface element of the object (area da), at a depth h. Then, let us consider the column of liquid over this surface element (Figure 5.11).

   If the liquid is in equilibrium, it means that the forces from the surrounding liquid over the column are equilibrated. This means that the surface element undergoes only the weight of the liquid column and an equal reaction of the object.

This weight is equal to:

$$P = Mg$$

where M is the mass and g the acceleration of gravity.

The mass M of the liquid column is linked to the volume and density ϱ by:

$$\varrho = \frac{M}{V}$$

The weight is then:

$$P = Mg = \varrho V g$$

And the volume is equal to:

$$V = h \cdot da$$

So that the weight is:

$$P = \varrho\, h\, da\, g$$

And the pressure applied on the object:

$$\mathcal{P} = \frac{F}{S} = \frac{P}{da} = \frac{\varrho\, h\, da\, g}{da} = \varrho h g$$

The buoyancy is equal to the sum of all pressure forces exerted by the liquid: it is upward, because the lower part of the immersed object receives bigger pressure than the upper part.

4. How to calculate the sum of all pressures, in particular for an object of complex shape?

   A: Let us call A this force, and P the weight of the immersed object (forget the P used before for the weight of the column of liquid; we now make a new calculation).

   Imagine that we wrap the immersed object in an infinitely thin and massless membrane (think of a very thin plastic bag that can become rigid when applied). How do weight and buoyancy change?

   This wrapping process does not change the weight and it does not change either the buoyancy *A:* remember that pressure acts on the surface, and that it does not depend on the material on which it acts.

5. Now imagine that, using a magical syringe, we take the substance of the object out of the membrane (without changing the shape of the membrane, which should not deform). How do weight and buoyancy change?

   A: Now the weight becomes nil, but the buoyancy does not change: it remains A.

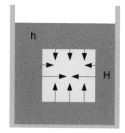

FIGURE 5.12 *Calculating pressure forces on a simple solid. The application of the strategy of making the problem simpler could lead the students on the path to another solution.*

6. Then, using another syringe, let us inject in the volume limited by the membrane a liquid of the same substance as the one in which the membrane is immersed. How do weight and buoyancy change?

   A: The weight changes (it is equal to the weight of the liquid having the same volume as the object), but the buoyancy does not change, again.

7. And now, let us remove the membrane. How do weight and buoyancy change?

   A: The liquid that was inside the membrane does not move, because it has the same density as the liquid outside: this means that forces acting on the liquid that was inside the membrane compensates: the weight of this liquid is equal to the buoyancy.

5.2.5.3.1 Another Calculation Showing Another Approach

The reasoning explained above calls for smartness, but could you get an idea of the question using another simple calculation?

Indeed, using the strategy proposed in all this document, of making the initial problem simpler, you could devise the following scheme, considering a simple shape (rectangular solid) in water (Figure 5.12).

On the two lateral edges, the pressure forces have a nil resultant, and the buoyancy is simply the vector sum of the forces acting on the upper part and on the bottom part.

The upper part is below the surface by a height h, so that the pressure is simply:

$$P_1 = \varrho_w \cdot g \cdot h$$

For the lower part, we have as well:

$$P_2 = \varrho_w \cdot g \cdot (H + h)$$

Moving from pressures to forces, with an area a for the object:

$$F_1 = P_1\, a = \varrho_w \cdot g \cdot h \cdot a$$
$$F_2 = P_2\, a = \varrho_w \cdot g \cdot (H + h) \cdot a$$

The sum of all pressure forces is:

$$A = F_2 - F_1 = \varrho_w \cdot g \cdot H \cdot a$$

With the two last factors $H \cdot a$, we recognize the volume V.

$$A = \varrho_w \cdot g \cdot V$$

And with this formula, we calculate the weight of the displaced volume of water.

5.3   HOW LONG SHOULD A FORK BE SO THAT WE CAN EAT IN THE DEVIL'S COMPANY?

5.3.1   An Experiment to Understand the Question

5.3.1.1   First, the General Idea

Of course, one will avoid dealing with the Devil (assuming its existence), but one can easily experiment with a metallic fork.

Here, the experiment will be simply to put a metallic spoon in boiling water, and measure the time after which you cannot keep the fork between the fingers (measuring the temperature will be interesting as well, because it gives an order of magnitude of what constitutes "hot", in the kitchen: what is a "hot" coffee, a hot bread, etc.

5.3.1.2   Questions of Methods

In order to explore experimentally the previous simple experiment, let us prepare a metallic fork, a wooden spoon, and a stick of potato (as long as possible). One after the other, you put one of their end in simmering water (try to get 95°C about, for example), and measure the amount of time that you can hold them, having your fingers at the opposite end to the one in the water.

You will observe that in some seconds, the metal is too hot to hold, whereas the wood remains cold, as well as the potato stick. You can try to make this observation quantitative by looking for thermal conductivity values in handbooks (online, probably today). For the potato, it is not sure that you will find the value, but one can approximate it with water because a potato is made of about 80% water) (Thybo et al., 2000); and water is not a good thermal conductor; if you find this assumption too elementary, calculate the barycentre of the values for water and for starch, assuming a 80/20 proportion.

5.3.1.3   Where the Beauty Lies

This experiment helps to understand why you have a sensation of coldness, when sitting on a metallic chair in the cool atmosphere of a house, in the summer (where the temperature is less than 37°C), but why you do not have this sensation when sitting on a wooden chair.

On wood, the heat from your body heats the chair, and the energy given to the wood remains under you. On the contrary, on a metallic chair, the heat flows rapidly around, so that your body gives more and more energy, without being able to heat the metal at the same temperature of 37°C as your body.

Now, a question: what do you predict that you will feel if you dip the end of a metallic spoon in a mixture of water and ice (mixing water and ice is a way to get the fixed temperature 0°C)?

5.3.2 The Question

Molecular and physical gastronomy (for short, molecular gastronomy) is the scientific discipline that deals with human nourishment, and cooking is normally followed by eating (This, 2009). In this field, also, questions can be found, such as the one found in an Alsatian proverb (in Shakespeare, and in Chaucer, you will find it also, but here the fork is instead a spoon) saying that even a very long fork is not enough to keep the Devil away when one shares a meal with it. The students are asked: "What do you think about it?"

Of course, it could appear strange to discuss a non-existing entity in science and technology, but the main goal of this book is to invite students to calculate, or to implement their calculation skills, and develop their creativity as well. Moreover, it is not forbidden (on the contrary: you don't have to be gloomy to be serious) to introduce some fun in scientific studies.

About this exercise, it is observed that it is obviously one for which many answers are possible, as explained below. This holds true for all the proposed questions of this book: again, the solutions given are only examples of possible calculations.

5.3.3 Analysis of the Question

5.3.3.1 The Data is Introduced

On the matter of fun in science, a story is told in physics laboratories about... we do not know who exactly. Some say that it was about Danish physicist Niels Bohr (1885–1962), as a student taking an exam. He was to use the "barometric gradient" (decrease in pressure with elevation) to calculate the height of the leaning tower of Pisa. Instead, Bohr proposed to use the barometer in various ways:

- to use the staircase and measure how many times the height of the barometer were needed to reach the summit;

- to attach the barometer to a long wire and measure the period of oscillation of the barometer in order to determine the length of the wire;

- to propose the gatekeeper of the tower give him a beautiful barometer in exchange for the needed information.

The diversity of solutions, and creativity in general, are discussed in a wonderful book by George Polya (1945).

Here, concerning the Devil, the issue is to invent the problem.

5.3.3.2 Qualitative Model

Problems like this one often upset students because it calls for more than the simple application of a formula; it calls for "inventing" the problem. However, we want to show that simple reasoning can help.

Let us begin by considering the Devil, assuming its existence for a while. If the question is asked about the length of a fork, it means that the Devil is to be kept far away. Why? In terms of physical chemistry, it could be because of:

- thermal questions,

- optical questions,

- electromagnetic questions,

- chemical questions,

- fluid dynamics questions.

There is no obvious reason for calculating fluid dynamics or chemistry, or electromagnetism, or optics. But everybody would admit that the Devil—living in Hell—should be very hot (and there are many popular tales in which is explained that the Devil's mark is burnt in wood), so thermal questions could be considered.

Of course, for such a problem, there are many possible solutions. Below, we give one of them.

5.3.3.3 Quantitative Model

Let T be the temperature of the Devil and L be the length of the fork needed to keep the Devil at the desired distance.

5.3.4 Solving

5.3.4.1 Looking for a Solving Strategy

There are many ways of solving the question, but if you consider the 14 definitions and formulas, you will see that one can use the Stefan formula of radiation of a black body (after all, the Devil is often depicted as black), or alternatively the heat conduction in the fork.

You could use the Stefan formula first, and then use the second Fourier equation as a validation (not done here but easily completed using *Maple* software).

5.3.4.2 Implementing the Strategy

Let us consider a hot body, radiating energy. The Stefan formula indicates that the power P radiated by a black body is proportional to the fourth power of the absolute temperature T, with a proportionality constant which is the product of the emitting area A and of the Stefan-Boltzmann constant:

$$P = \varepsilon \cdot \sigma \cdot A \cdot T^4$$

Here, ε is the emissivity, between 0 and 1, taking into account the incomplete blackness of the body. For simplicity, we shall take it equal to 1.

This power corresponds to an emitted energy E during the time t:

$$E = Pt = \sigma \cdot A \cdot T^4 \cdot t.$$

This energy E is radiated radially. When it travels a distance r, it distributes on a sphere having this radius, and whose area S is equal to:

$$S = 4\pi r^2$$

At this point, we have a choice: either we decide to do the next step rapidly, with the risk of making mistakes, or we can be slow (my proposal) and try to be sure of what we do. By experience, even I can be fast, I prefer the second option. Let e be the quantity of energy received by an object (our body) of cross section a at the distance r (during the time t): it is a proportion, with a ratio $\frac{a}{S}$:

$$e = \frac{a}{4\pi r^2} \cdot E = \frac{a}{4\pi r^2} \cdot \sigma A T^4 t$$

About the rest of the calculation, I propose to take a moment to discuss the question of proportionality, because, from experience, I know that it is the cause of many errors.

Let's not be snobbish, even if we all feel that at university we know how to do such a calculation: knowing the energy for an area A, how can we determine the energy for an area a if there is proportionality?

Some people express the ratios as fractions, make the two fractions equal to each other, perform cross-multiplications, and solve the resulting equation. Others are faster... but after years of having trainees in the lab, I can assure you that it's worth going slowly: there have been so many mistakes! And then, as much as one can make a mistake without serious consequences for an exam at university, it is better to be sure of oneself, in professional life : imagine that you build a bridge, that you dose an active ingredient in a drug!

Could not we say slowly: the sphere of radius r, and of area S receives an energy E; because of proportionality, an energy S times smaller (i.e., $S/S = 1$) would get an energy S times smaller, i.e., E/S. An area a, i.e., a times larger than 1, would get an energy a times larger than an area 1. Thus, the energy received by the area a is $a \times E/S$. Safer, is it not?

This being written, the body receiving this energy is heated so that if no energy is evacuated and its temperature is increased by ΔT given by:

$$e = m c_p \Delta T$$

Here, there is an assumption: the heat capacity is supposed to be constant with temperature.

We can now write that the energy received is for heating the body:

$$\frac{a}{4\pi r^2}\sigma A T^4 t = m c_p \Delta T$$

In this equation, we look for the distance r:

$$r = \sqrt{\frac{a\sigma AT^4 t}{4\pi m c_p \Delta T}}$$

(if you want to be sure of your result, you can check it using the software)

5.3.5 Expressing the Results

5.3.5.1 The Formal Result is Written Down Again

We found this expression, for the distance r:

$$r = \sqrt{\frac{a\sigma AT^4 t}{4\pi m c_p \Delta T}}$$

What does it mean? It is the distance for which a temperature increase ΔT is obtained after a time t. The time can be the duration of the meal, and the temperature increase is the one that our body can bear.

5.3.5.2 Finding Digital Data

The Stefan-Boltzmann constant is known: in SI units, it is $5.6703 \ 10^{-8}$ W $\cdot$ m^{-2} $\cdot$ K (NIST, 2020).

For the size of the Devil, let us assume that, if we eat with him, its size is slightly taller, wider and thicker than us ($2 \times 1 \times 1$). For the temperature of the Devil, we can observe that the Devil is not human, and it can be assumed that it is hotter than anything human. Humans can produce very high temperatures in solar furnaces, or with plasma torches, and, more recently in Tokamaks, for producing nuclear fusion, i.e., 150 million kelvins (Futurism, 2022). But, since the Devil was already here before the Big Bang, we can decide for an even greater number, such as 1000 billion kelvins.

The time t can be chosen as the duration of a meal, i.e., 3600 s.

For the cross section of the body, we can assume 2 m high and 0.5 m wide.

Let assume a body of mass $m = 80$ kg.

The heat capacity is assumed to be the one of water. We know that 1 cal is the energy that heats 1 g of water by 1°C. It is equal to 4.18 J. So that c_p is equal to 4.18 kJ/kg. K.

Now, for the determination of ΔT, we can observe that our body is made of water and proteins, as "meat", and we would coagulate at too high a temperature (we could decide that this threshold is 40, or 50°C, for example). This fixes ΔT.

5.3.5.3 Introduction of Data in the Formal Solution

$$evalf \left(subs \left(a = 2 \cdot 0.5, \sigma = 5.6703e - 8, A = 10, \right.\right.$$

$$\left.\left. T = 1e12, t = 3600, m = 80, c_p = 4.18, \Delta T = 40, r = \sqrt{\frac{a\sigma AT^4 t}{4\pi m c_p \Delta T}} \right)\right)$$

$$r = 1.10 \times 10^{20}$$

Here again, the number of digits is clearly non-significant, and only an order of magnitude is to be considered: 10^{20} !

The units as SI ones, *i.e.*, m.

5.3.6 Conclusions and Perspectives

5.3.6.1 Conclusions

The radius (in m) being much more that the size of the Earth, one can see that the Devil "is" indeed so hot that we could not survive on a very large table in its company: the proverb is validated by our calculation.

More seriously, this exercise is an opportunity to train students to behave as scientists, asking "how much?" about everything. Indeed, these questions are like "jogging".

5.3.6.2 Perspectives

We observed that creativity is needed for problems, and we could also have tackled the question by calculating the question by considering the temperature increase of the fork due to heat conduction. Here the Fourier model can be used, with the thermal conductivity found (Lide, 2005) to be $80.2 \text{ W} \cdot \text{cm}^{-1} \text{ K}^{-1}$. The boundary conditions would be:

- the temperature of the end of the fork on Devil's side is 10^{12} K.

- the temperature of the other end is at room temperature.

- after 1 h, the temperature in our hand is 55°C.

And of course, when the formal calculation has led to a solution, one has to check it.

5.4 HOW MUCH ACIDITY IS PRODUCED WHEN A BOTTLE OF WINE CONTAINS AIR?

5.4.1 An Experiment to Understand the Question

5.4.1.1 First, the General Idea

It is interesting, once in a lifetime, to make wine and vinegar: it will show you probably how much better good professionals are than you!

Start from grapes, that you press and store them along with the juice in a vessel. After some days or weeks (depending on the temperature and other conditions), you will observe bubbles in the liquid (carbon dioxide, CO_2), and, if you drink it, you will perceive some "alcoholic" sensation: sugars (mainly D-glucose) are fermented by yeasts into (in particular) ethanol (CH_3CH_2OH).

Here, a remark : above, I gave the formula of ethanol, and, personally, it seems useless to me, as I have known well that the formula of ethanol is CH_3CH_2OH, since I was a child (I "fell for chemistry" at the age of six, when I received a box of chemistry for Christmas). However, I also know from experience that some younger friends do not have this intimate knowledge, and it is to them that I am thinking by forcing myself to a little redundancy. For those who do know, I do not think that it is completely useless to come back to it because it gives one the opportunity to ask oneself questions in order to improve one's knowledge. For example, is this expression, "CH_3CH_2OH" an elementary formula? a developed one? a semi-developed one? and do you know enough about the International Chemical Identifier, or InChI, which is another way to describe molecules (McNaught, 2006)? And what about the "InChI-Key"? Or about the "CAS Number"? The PubChem CID? The Canonical SMILES? You see, even when we "know", we do not know enough, and any information that we think we have is probably an invitation to go further. Finally, let us decide that, for this question only, I shall write always the formula after the name, but otherwise, we will be more subtle.

Coming back to our question, once you have made "wine", leave it at the open air for more time and you will perhaps observe soon (*i.e.*, after some days or weeks, depending on the temperature and exposition) that small flies over the liquid. If you drink it, you will now perceive that acidity developed: wine turned into vinegar, when ethanol (CH_3CH_2OH) was oxidized into acetic acid (CH_3COOH) by various micro-organisms, such as *Mycoderma acetii*.

5.4.1.2 Questions of Methods

Ethanolic fermentation was the cause of much scientific debate, at the time of the French microbiologist Louis Pasteur (1822–1895) and the German chemist Justus von Liebig (1803–1873) because enzymes and micro-organisms were hardly known, and chemistry as well was not clear. Finally, it was shown that yeasts contain enzymes that transform sugars from grapes (D-glucose, D-fructose, sucrose) into ethanol (CH_3CH_2OH).

For acetification, ethanol (CH_3CH_2OH) being transformed into acetic acid (CH_3COOH), the reaction corresponds to an oxidation (Clayden *et al.,* 2001). And indeed, you can also obtain acetic acid (CH_3COOH) from ethanol (CH_3CH_2OH) by various chemical, non-enzymatic, processes, for example using potassium permanganate. Potassium permanganate? It is found as dark purple crystals, and one of them in some grams of water is enough to give to the liquid a beautiful purple color (the color of emperors, as was said in the past). The formula is $KMnO_4$. And for the experiment, you simply add a dilute solution of potassium permanganate $KMnO_4$ to ethanol (CH_3CH_2OH) (you can make it acid using dilute sulfuric acid, H_2SO_4), and you see the purple color turning brown. The equation of the chemical transformation can be written:

$$CH_3CH_2OH + O_2 \rightarrow CH_3COOH + H_2O.$$

Look: potassium permanganate $KMnO_4$ does not appear, but only dioxygen O_2 is present in the equation, and the reason is that I wanted to describe both transformations, by micro-organisms or by oxidizers. Would you be able to write the chemical equation, in the case of potassium permanganate?

Also you are perhaps more familiar with the word "oxygen" than with "dioxygen", as I was myself in the past, but I propose to change our ways for the better: oxygen

(O) is the element, and dioxygen (O_2) is a molecular species. Keeping the same word for the two would create confusion (Myers, 2012). This seems a detail, but I can tell you that there are many people outside chemistry that do not understand the difference between a compound (for example, ethanol) and a molecule: a compound is a particular category of molecules, and a molecule is a tiny object made of atoms (This vo Kientza, 2021d). The confusion can be terrible: I remember a science journalist (for sure, this one in particular was a bad one) who thought that 450 molecules (yes, the assemblies of atoms) were responsible for the odor of wines, and he was upset when I told him that there were hundreds of billions of billions of molecules for each odorant compound (Roullier-Gall *et al.*, 2014).

5.4.1.3 *Where the Beauty Lies*

The transformation that we considered is the key to a good wine vinegar (with micro-organisms, not with potassium permanganate!), for which there is a frequent misconception, *i.e.*, a "mother of vinegar" should be inside the liquid.

"Mother of vinegar" is the name given to a veil that can be observed at the upper surface of some vinegars, or inside it; in the second cases, it is the sign of a bad practice.

Indeed, when wine is left in the presence of air, micro-organisms of the species *Mycoderma acetii* and others (Mas *et al.*, 2014), initially present in the air, develop at the surface of wine, because they find here convenient conditions to do so: they use the dioxygen O_2 of air and the ethanol CH_3CH_2OH of wine to create acetic acid CII_3COOH.

But when users are not cautious or when they do not know enough about the process, they sometimes add more fresh wine to the already transformed liquid, and this puts the veil of micro-organisms into the liquid. The bad thing, then, is that the micro-organisms are deprived of dioxygen, and now consume the acetic acid that had formed, instead of creating it, and the flavor of the liquid turns bad.

So, contrary to a popular idea, do not consider the mother of vinegar from friends as a precious gift, and, instead, wait for some days, so that the micro-organisms of the environment come spontaneously in your open vessel, at the edge of the window for example.

5.4.2 The Question

Let us have a bottle of wine. The liquid does not fill the bottle entirely, and some air remains between the liquid and the cork. But air contains dioxygen, which can turn ethanol into acetic acid, if the right micro-organisms are present.

How much acidity can be produced?

5.4.3 Analysis of the Question

5.4.3.1 *The Data is Introduced*

As we now know, we introduce the data, but we also translate the units into SI units: a bottle of wine has a capacity of 0.75 L, *i.e.*, 7.5×10^{-4} m³.

The *pKa* of acetic acid CH_3COOH is equal to 4.76 at 25°C (Lide, 2005).

FIGURE 5.13 *Gas in a bottle. This picture has the same goal as the previous ones. The fact that it is simple is also a demonstration of the fact that the exercises from this chapter are simple.*

5.4.3.2 Qualitative Model

The question discussed here is a trap, as will be shown later, and we make the calculation below only because it is one that students sometimes fail to make, by lack of method.

Let us begin with a picture, as we always must (Figure 5.13).

Here we simply consider the wine and air.

5.4.3.3 Quantitative Model

We introduce the parameters that describe the picture:

- V: total volume of the bottle.
- v: volume of air over wine, in the bottle.
- p: percentage of ethanol (CH_3CH_2OH) in wine.

If the conditions are convenient, the dioxygen (O_2) of the air is used for the oxidation of ethanol (CH_3CH_2OH) into acetic acid (CH_3COOH) according to the general equation:

$$CH_3CH_2OH + O_2 \rightarrow CH_3COOH + H_2O$$

Of course, the reaction is limited by the reactants: here dioxygen (O_2) is clearly the limiting one.

Now, from the equation above, we see that the oxidation of 1 mole of ethanol (CH_3CH_2OH) generates 1 mole of acetic acid (CH_3COOH).

Of course, wine will not get its pH from this acetic acid (CH_3COOH) only, because it contains a lot of other acids (*e.g.*, tartaric, malic, succinic), but if we want to make a simple calculation we will first assume that we calculate only the effect of acetic acid (CH_3COOH) on a liquid that would be initially neutral with no buffer.

Here, an observation about writing pH: you see that because it is a quantity, it should be in italics, as we said before; note that the international community decided to write in roman letters pH (Nornby, 2000; Jensen, 2004).

5.4.4 Solving

5.4.4.1 Looking for a Solving Strategy

From the volume of air, we can deduce the volume of dioxygen (O_2).

From this volume, we will determine the number of moles of dioxygen (O_2), that is the number of moles of acetic acid (CH_3COOH) produced.

From this number of moles, we can determine the pH of an aqueous solution.

5.4.4.2 Implementing the Strategy

Let assume that the air in the bottle occupies a cylinder of internal diameter d and length L. Its volume is:

$$v = \frac{\pi \cdot d^2}{4} \cdot L$$

In this volume, only dioxygen (O_2) contributes to the oxidation of ethanol (CH_3CH_2OH).

Let us then determine how much dioxygen (O_2) is present. In the air over the wine, the partial pressure of dioxygen (O_2) in air is P_{O_2}.

Having a volume and a pressure, we guess that we have to apply the formula for ideal gases:

$$P_{O_2} \cdot v = nRT$$

In this relationship, we see that the only unknown quantity is the number of moles n. It can be calculated as:

$$n = \frac{P_{O_2} \cdot v}{RT} = \frac{P_{O_2} \cdot \left(\frac{\pi \cdot d^2}{4} \cdot L\right)}{RT}$$

If acetic acid (as said, characterized by its pK_A) was alone in solution in water (a simplification, as there are indeed many other acids), then the pH would decrease from 7 to a value that can be calculated using a formula that you have certainly learned, for weak acids:

$$\mathrm{pH} = \frac{1}{2}(pK_A - \log 10(c)),$$

where c is the concentration (in moles per liter of liquid). Here we use the *Maple* function log10, for decimal (or base 10, or "common") logarithm; it could also be written log[10], or $\log_{10}$.

Let us observe here that later, we shall use software for the calculation of the pH without having to remember such equations.

Then

$$\mathrm{pH} = \frac{1}{2}\left(pK_A - \log 10\left(\frac{\left(\frac{P_{O_2}\cdot\left(\frac{\pi\cdot d^2}{4}\cdot L\right)}{RT}\right)}{V - \frac{\pi\cdot d^2}{4}\cdot L}\right)\right)$$

5.4.5 Expressing the Results

5.4.5.1 The Formal Result Found is Written Down Again

$$\mathrm{pH} = \frac{1}{2}\left(pK_A - \log 10\left(\frac{\left(\frac{P_{O2}\cdot\left(\frac{\pi\cdot d^2}{4}\cdot L\right)}{RT}\right)}{V - \frac{\pi\cdot d^2}{4}\cdot L}\right)\right)$$

5.4.5.2 Finding Digital Data

The partial pressure of dioxygen (O_2) in air is about 0.2 atm, *i.e.,* 0.2×10^5 Pa.
For the length of the air cylinder, we fix $L = 0.05$ m.
For the diameter, we choose: $d = 0.01$ m.
For the volume V, we have 0.75 L.

5.4.5.3 Introduction of Data in the Formal Solution

$$evalf\left(subs\left(pK_A = 4.76, P_{O_2} = 0.2e5, L = 5e-2, R = 8.32,\right.\right.$$

$$\left.\left. T = 300, d = 0.01, V = 0.75, pH = \frac{1}{2}\left(pK_A - \log 10\left(\frac{\left(\frac{P_{O_2}\cdot\left(\frac{\pi\cdot d^2}{4}\cdot L\right)}{RT}\right)}{V - \frac{\pi\cdot d^2}{4}\cdot L}\right)\right)\right)\right)$$

$$pH = 4.57$$

5.4.5.4 Discussion: Playing with Parameters in Order to Explore the Solution Space

With this result, corresponding to an ethanolic solution being acetified, we find a value for pH that is slightly higher than of wine (3–4). How is it possible? Before answering, let us discuss the question of "playing with parameters for the exploration of the solution space", because it is sometimes considered difficult by students. Here, this would mean -for example- looking for extreme values.

For example, here, one could calculate the pH for two extreme cases, *i.e.*, no air at all, on one hand, and all ethanol (CH_3CH_2OH) being transformed into acetic acid (CH_3COOH) on the other hand.

For no air, of course, the pH would remain equal to 7, with our assumptions.

If all ethanol (about 10%) were transformed into acetic acid, this would mean that a mass of about 100 g for 1 L of water would be oxidized.

This would mean about 2 moles (46 g for ethanol) of acetic acid, so that the pH would become:

$$pH = evalf\left(\frac{1}{2} \cdot \left(4.76 - \log 10\left(\frac{2}{1}\right)\right)\right)$$
$$pH = 2.23$$

This would be more acidic than wine... and equal to the pH of many vinegars, or of lemon juice.

5.4.6 Conclusions and Perspectives

5.4.6.1 Conclusions

In the beginning of this section, we announced that the calculations above are flawed in various ways.

1. The assumptions that we made are clearly wrong, in particular because food ingredients contain a lot of acids and bases, with possible buffers (Pietry and Land, 2020) that oppose pH changes. Indeed, we calculated the acidification starting from pure water, but if wine has a lower pH than the one calculated, the pH has to decrease. This acid pH of wine is due to various organic acids, such as (+)-tartaric acid ((2R, 3R)-dihydroxybutanedioic acid, HOOC-CHOH-CHOH-COOH), L-malic acid ((L)-2-hydroxybutanedioic acid, HOOC-CH$_2$-CHOH-COOH), citric acid (2-hydroxypropane-1,2,3-tricarboxylic acid,($CH_2CO_2H)_2$), that are present in grapes with their salts (This, 2021). During wine making, other acids can play a role: acetic acid (see above), butyric acid (butanoic acid, $CH_3CH_2CH_2CO_2H$), lactic acid ((L)-2-hydroxypropanoic acid, $CH_3CHOHCOOH$) and succinic acid (butanedioic acid, $(CH_2)_2(CO_2H)_2$).

2. We simply applied the formula for the pH of weak acids pH $= \frac{1}{2}(pKa - \log 10(c))$, but what if we did not remember this formula? Instead of reproducing the courses in solution chemistry, we propose to move to the digital era, *i.e.*, to use software for formal calculation such as *Maple;* we shall do this later, in the question about sulfur dioxide in wine, but it could apply here as well.

5.4.6.2 *Perspectives*

Discussing the assumptions in a way improves a solution. Now, and now only, we could decide to make a more complex pH calculation, taking into account the various mineral species contributing to the pH of wine. In particular, we could investigate solutions containing (+)-tartaric acid and one of his salts, fixing the proportions of the two so that the solution considered would have the pH of usual wines. The game is endless.

5.5 WHY YOU SHOULD NEVER POUR WATER ON FLAMING OIL?

5.5.1 An Experiment to Understand the Question

5.5.1.1 *First, the General Idea*

Here, we shall certainly NOT carry out the experiment! Perhaps you heard before that one should NOT pour water on flaming oil, and this bears repeating: one should NOT pour water on flaming oil, because water would fall to the bottom of the pan, and its sudden vaporization would make an explosion, projecting flaming oil all around (CFPA, 2012).

However, it can be a good practical laboratory exercise to invite students to stop a fire in burning oil. Of course, also for this activity, the students should be alerted on the way to do it: outside, in a very open space, with a lot of precautions (Mendham, 2022).

5.5.1.2 *Questions of Methods*

Phenomena including fire were investigated by many scientists of the past, in particular by the wonderful English chemist Michael Faraday.

Having lost his father when young and being from a very poor family, he later—when he was the director of the Royal Institution of London—decided to organize free scientific lectures for the orphans of London. These "Christmas lectures" still exist today. One of the most popular ones was turned into a book, *A course of six lectures, adapted to a juvenile auditory, on the chemical history of a candle* (Faraday, 1861). In it, Faraday explains many of the experiments that you can make with a simple candle. One is to burn a candle in a beaker (the wall of the beaker has to be higher than the candle), so that the carbon dioxide CO_2 created by the burning of the candle accumulates progressively at the bottom of the beaker and finally the candle stops burning.

Why? Remember this simple formula: $d = M/29$, where d is the density of a gas compared to air, and M the molar mass of the gas (do you know where this formula comes from?). With carbon dioxide, M is equal to 44 ($12 + 16 + 16$), which is clearly denser than for air (about 29): the carbon dioxide accumulates, moving air (and dioxygen in particular) upward, showing that combustion needs it to occur. The same holds with flaming oil.

5.5.1.3 Where the Beauty Lies

In order to stop an oil fire safely, it is good to prepare a small old pan with a small quantity (a layer of about 1 cm) of ordinary oil and a larger pan at hand. No surface (ceiling) should be too close to the place where the experiment is done (do the experiment outside, far away from all material that could burn, such as plastic, gasoline, etc.).

First heat the oil in the small pan until it burns, and try to stop the fire by first stop the heating and second by calmly putting the larger pan over the small one: the dioxygen enclosed in the small pan will be consumed in some seconds, and the fire will be extinguished.

The issue is "how many seconds ?". If you make first the experiment with the larger pan over the small one for only two seconds, you will likely see that the oil still burns when you take the large pan away. Repeat putting the large pan over the small one, but calmly, giving you time to turn down the heat: after some tens of seconds, you will see that the fire is indeed quenched.

On the other hand, avoid absolutely to make the experiment of adding water to the flaming oil, as the following calculation shows.

5.5.2 The Question

And now the question has to be repeated: why is it extremely dangerous to add water to flaming oil?

5.5.3 Analysis of the Question

5.5.3.1 The Data is Introduced

Here, no data is given.

5.5.3.2 Qualitative Model

Let assume that oil burns. How to characterize it? Obviously with the temperature. But what is the temperature of flaming oil? Culinary devices for frying are often limited at 170°C for many health reasons, but having a limit because oils can catch fire when the temperature in particular oil can catch fire when the temperature is higher than the smoking point, *i.e.*, 190°C for animal fat, and 230°C for vegetable oils (Eveleigh, 2021).

If water were added (don't do it!), it would sink under the oil because its density is higher than the density of oil. Being in an environment where the temperature is much hotter than the boiling point (100°C), water would evaporate rapidly. The resulting fast increase in volume would probably cause an explosion: steam occupies more volume....

However, steam occupies more volume than liquid water, so that the flaming oil is ejected in all directions.

You are invited to look at https://www.youtube.com/watch?v=U4j9Abq_bR0

5.5.3.3 Quantitative Model

Here, there is no real need to look for a resolution strategy, because the "qualitative model" solves the question. But let us insist: the calculation "backbone" that we use here is for training, and the goal is that it becomes so well internalized that it does not need any longer to be explicitly used.

Also, up to now, we used formal calculation, but sometimes the calculation of orders of magnitudes is enough, as here.

Let us assume that 1 L of water were added: this would mean 1 kg, or about 50 moles ($\sim 1000/18$) of water.

This water would fall at the bottom of the pan containing oil (its density is 1000 kg/m^3, compared to about 900 for oils)(Lide, 2005), and it would evaporate very rapidly because the temperature would be higher than the boiling temperature of water.

Before moving to the end of the calculation, let us stop at this "very rapidly", because it is made of an adverb and an adjective. And it is not good science to use such words: when you see adverbs or adjectives in sentences that we write, immediately ask yourself "How much?" and replace these words by the answer to the question.

Yes, how fast does water evaporate? Could you guess how to calculate it? We shall not do it now, but you see that the game of calculation is a never-ending one.

Let us go back to our calculation. We know that one mole of water makes 24 L of steam at 20°C, but here, it would be better to know the volume of one mole at 100°C. This can be done using the formula for ideal gases, at the two temperatures 20°C and 100°C :

$$P_1 \cdot V_1 = RT_1$$
$$P_2 \cdot V_2 = RT_2,$$

As the pressure is the same, this would give:

$$V_2 = \frac{V_1 \cdot T_2}{T_1}$$

And the result is:

$$evalf \left(subs \left(V_1 = 24, T_2 = 373, T_1 = 293, V_1 \frac{T_2}{T_1} \right) \right)$$

30.55

Now we can use this to calculate that the volume of water would generate rapidly about 50 times that, i.e., about 1.5 m^3.

Suddenly, this steam would project burning oil everywhere, with an explosion (a sudden variation of pressure).

5.5.4 Conclusions and Perspectives

If there is a fire in a kitchen, stay calm, extinguish the fire with a material that does not burn and that you can put over the flame.

REFERENCES

AOAC International. 2019. *Official Methods of Analysis of AOAC International*, 21rst edition.

This is very important information: when you need a chemical method, look at this first because it was created as a common tool for the industry.

Belitz HD, Grosch W, Schieberle P. 2004. *Food chemistry* (3rd ed), Springer Verlag, Heidelberg, Germany.

This is the second time that I provide this reference... and not the last one.

BIPM. 2012. *Evaluation of measurement data. The role of measurement uncertainty in conformity assessment*, JCGM 106:2012. © JCGM 2012.

BIPM. 2022. *SI Brochure: The International System of Units (SI)*, https://www.bipm.org/en/publications/si-brochure, last access 2022-01-04.

Idem as for Food Chemistry.

Bocuse P. 2012. *The complete recipes*, Flammarion, Paris, France.

Bonnefons N. 1655. *Les délices de la campagne*, Raphael Smith, Amsterdam, The Netherlands.

CFPA. 2012. *Fire safety in restaurants*, CEPA-E Guideline No 9:2012 F, http://cfpa-e.eu/wp-content/uploads/files/guidelines/CFPA_E_Guideline_No_9_2012_F.pdf, last access 2022-01-06.

Champion D, Hervet H, Blond G, Le Meste M, Simatos D. 1997. Translational diffusion in sucrose solutions in the vicinity of their glass transition temperature, *The Journal of Physical Chemistry B'*, 101, 10674–10679.

Chaucer G. 1386. *The Canterbury tales*, https://quod.lib.umich.edu/c/cme/CT?rgn=main;view=fulltext, last access 2022-01-12.

Clayden J, Greeves N, Warren S. 2012. *Organic chemistry* (2nd ed), Oxford University Press, Oxford, UK.

There are many textbooks for organic chemistry, and some are better than others, but also some are better for us personally than others. If one of them does not suit you (including this one), change it.

Cultures Sucre. 2020. *La boîte à questions*, https://www.lesucre.com/.

Eveleigh L. 2021. *Oxidation of dietary lipids*, in Burke R, Kelly A, Lavelle C, This vo Kientza H (eds), *Handbook of molecular gastronomy*, CRC Press, Boca Raton, USA, 305–310.

Faraday M. 1861. *A course of six lectures, adapted to a juvenile auditory, on the chemical history of a candle*, https://gallica.bnf.fr/ark:/12148/bpt6k778489.image, last access 2022-01-07.

Feynman R, Leighton R, Sands M. 1963. *The Feynman lectures on physics*, The California Institute of Technology.

A lot of physicists were trained with this series of books, and there are good reasons for it. Read them slowly.

Futurism. 2022. *What's the highest known temperature?*, https://futurism.com/what-is-the-highest-known-temperature-2, last access 2022-01-06.

Hardy. 1998. *ITS-90 formulations for vapor pressure, frostpoint temperature, dewpoint temperature and enhancement factors in the range -100 to +100° C.* http://www.meteo.tomkii.net/humidity/dewpoint_its90formulas.pdf, last access 2018-11-15.

Isengard HD. 2018 . *Méthodes de détermination de la teneur en eau dans les produits riches en sucre. Eau, sucre et poudres alimentaires*, https://www.sucre-info.com/content/uploads/2002/01/isengard.pdf, last access 2018-10-12.

IUPAC. 2018. *Colloidal*, https://goldbook.iupac.org/html/C/C01172.html, last access 2018-11-15.

Jensen WB. 2004. The symbol for pH, *Journal of Chemical Education*, 81(1), 20–21.

Juvin J, Belhomme MC, Castex S, Bliard C, Haudrechy A. 2019. *Un bon croquis vaut mieux qu'un long discours?*, Notes Académiques de l'Académie d'Agriculture de France, 2, 1–9.

A very important article if you learn chemistry. Do not hesitate to translate it into your own language.

Lepetit J. 2008. Collagen contribution to meat toughness: theoretical aspects, *Meat Science*, 80(4), 960–967.

Lide DR (ed.). 2005. *CRC Handbook of chemistry and physics*, Internet Version 2005, <http://www.hbcpnetbase.com>, CRC Press, Boca Raton, FL.

Mas A, Torija MJ, Garcia-Parrilla MdC, Troncoso AM. 2014. Acetic acid bacteria and the production and quality of wine vinegar, *The Scientific World Journal*, 2014, 394671, 1–6.

McNaught A. 2006 *The IUPAC International Chemical Identifier: InChl*, Chemistry International, IUPAC, 28(6), https://publications.iupac.org/ci/2006/2806/2806-pp12-15.pdf, last access 2022-01-06.

If you look at the options of your chemistry software (Marvin, Avogadro, etc.), you will see that there are many possibilities, such as XYZ, SMILE, MDL, etc.

McQuarrie DA, Simon JD. 1997. *Physical chemistry*, University Science Books, Sausalito, California.

I tried many textbooks in the past but I love this one.

Mendham. 2012. *Kitchen safety: grease fires*, https://mendhamfd.org/kitchen-safety-grease-fires/, last access 2022-01-07.

Myers RJ. 2012. What are elements and compounds, *Journal of Chemical Education*, 89, 832–833.

NIST. 2020. *Codata value: sigma*, https://physics.nist.gov/cgi-bin/cuu/Value?sigma.

Park KC, Pyo HS, Kim WS, Yoon KS. 2017. Effects of cooking methods and tea marinades on the formation ofbenzo[a]pyrene in grilled pork belly (Samgyeopsal), *Meat Science*, 129, 1–8.

Pietri J, Land D. 2020. *Introduction to buffers*, Chemistry-LibreTexts, https://chem.libretexts.org/@go/page/78249, last access 2022-01-07.

Polya G. 1945. *How to solve it*, Princeton University Press, Princeton, NJ.

Roullier-Gall C, Boutegrabet L, Gougeon RD, Schmitt-Kopplin P. 2014. *A grape and wine chemodiversity comparison of different appellations in Burgundy: Vintage vs terroir effects.* Food Chemistry 152, 100–107.

Slade L, Levine H, Ievolella J, Wang M. 1993. The glassy state phenomenon in applications for the food industry: Application of the food polymer science approach to structure–function relationships of sucrose in cookie and cracker systems, *Journal of the Science of Food and Agriculture*, 63, 133–176.

Starkzak ME. 2010. *Energy and entropy*, Springer, Heidelberg, Germany.

Thybo AK, Bechmann IE, Martens M, Engelsen SB. 2000. Prediction of sensory texture of cooked potatoes using uniaxial compression, near infrared spectroscopy and low field 1H NMR spectroscopy, *Lebensmittel Wissenschaft und Technologie*, 33, 103–111.

This H. 2009. *Molecular gastronomy, a chemical look to cooking*, Accounts of Chemical Research, 42 (5), 575–583.

This H. 2010. *Cours de gastronomie moléculaire N°2: les précisions culinaires*, Quae/Belin, Paris.

This H. 2021. Des cristaux d'Auguste Laurent et des techniques d'analyse optique de Jean-Baptiste Biot furent directement à l'origine de la découverte de la chiralité par Louis Pasteur, *Notes Académiques de l'Académie* d'agriculture de France, 9, 1–33.

This vo Kientza H. 2021a. *Dispersed Systems Formalism (DSF)*, in Burke R, Kelly A, Lavelle C, This vo Kientza H (eds), *Handbook of molecular gastronomy*, CRC Press, Boca Raton, USA, 207–212.

If you study physical chemistry, or formulation, or food, this DSF tool will be very helpful.

This vo Kientza H. 2021a. The cousins of whipped cream: "Chantillys", in Burke R, Kelly A, Lavelle C, This vo Kientza H (eds), *Handbook of molecular gastronomy,* CRC Press, Boca Raton, USA, 105–106.

This vo Kientza H. 2021b. Experimental flavour workshops, in Burke R, Kelly A, Lavelle C, This vo Kientza H (eds), *Handbook of molecular gastronomy,* CRC Press, Boca Raton, USA, 635–642.

This vo Kientza H. 2021c. Expansion, in Burke R, Kelly A, Lavelle C, This vo Kientza H (eds), *Handbook of molecular gastronomy,* CRC Press, Boca Raton, USA, 291–294.

This vo Kientza H. 2021d. The right words for improving communication in food science, food technology, and between food science and technology and a broader audience: *Handbook of molecular gastronomy,* CRC Press, Boca Raton, USA, 626–634.

CHAPTER 6

Liquids (L)

$\mathbf{A}$S AN INTRODUCTION to the previous chapter, we discussed the organization of the chapters, based on increasing DSF complexity. The symbol G, for gas, is before O, for oil, or S for solids, or W, for water, which are the most frequent phases in molecular and physical gastronomy. But here you see neither O nor W, but L: remember that there is a game, in sciences of nature, to represent unknown quantities or objects by letters. If we want to group "oil" and "water" phases, we can admit that they are liquids, hence the L.

6.1 THE DISTANCE BETWEEN TWO AIR MOLECULES WAS FORMERLY CALCULATED, BUT WHAT ABOUT THE DISTANCE BETWEEN TWO WATER MOLECULES?

6.1.1 An Experiment to Understand the Question

6.1.1.1 First, the General Idea

1. Let us use an empty aluminium can, and affix a straw with glue, so that the seal is water proof.

2. Then use a pen to make a horizontal line on the straw, some centimeters over the level of the can.

3. Put the can in the fridge for about 10 minutes.

4. Then fill it in with cold water from the fridge up to the level of the line (you can use a syringe to inject the liquid in the straw).

5. Put the whole system on a radiator in the winter, or in a hot room in the summer.

6. After 10 minutes, observe the level of water in the straw. Make a line at this level, on the straw.

7. Put the system back in the fridge for about 10 minutes and observe again the level.

8. If possible, measure the temperature in the fridge and in the hot room.

DOI: 10.1201/9781003298151-6

6.1.1.2 Questions of Methods

The previous experiments shows that water contracts when it is cooled, and expands when it is heated (we went fast about the anomalous dilatation around the temperature of 4°C, but of course, you need to be aware of it). We could have added a measurement of the mass, using a precision balance, and we would have observed that the mass does not change, from the fridge to the hot room and back. This is why, in laboratories, it is a much better practice to measure masses instead of volumes.

By the way, on precision burettes, the very thin line that is carved in the restricted top part is indicating a certain volume but only at a certain temperature (generally, this kind of burettes is sold along with a certificate where this information is given). But one can understand that precision can be obtained only if one works in a temperature-controlled room (Harris, 2007).

Measuring masses has many other advantages, as can be seen when using capillary tubes to deposit precise quantities of liquids on plates used for thin layer chromatography (TLC) (Valverde *et al.*, 2007). Here, the idea is to deposit a drop of a solution of several compounds on a plate on which a powder (often silica gel, or aluminium oxide, or cellulose) is deposited; when the plate is standing in a small quantity of solvent, the migration of the solvent separates the various solutes, because they have different affinities for the solvent and the powder.

TLC can become quantitative if the quantity of solution that one deposits is known precisely, but one can observe that even when brand new capillary tubes (*e.g.*, 20 μL) are put in very pure water, the level is not the same in all tubes because the glass is rapidly contaminated by the environment, changing the surface tension. On the other hand, if you use (correctly!) a precision scale (*e.g.*, with which you can detect variations of 10^{-5} g), the mass of deposited liquid can be known with the precision of the scale! And here is the trick: you weigh the empty capillary tube, you weigh the tube after filling, you deposit the solution, and you weigh again; by subtraction, you can know the mass of solution deposited on the plate.

Let us repeat: measuring masses is better practice than measuring volumes!

6.1.1.3 Where the Beauty Lies

Let us now calculate the uncertainty of various methods, when making a solution containing 100 mg of a compound C in 100 mL of a solvent S. We shall assume that the molar mass of C is 100 g.

1. First we use a balance to weigh the compound C, and a burette for the solvent. The final concentration (in mol/L) c is:

$$c_v = \frac{n}{V}$$

with n the number of moles of C, and V the volume.

Because the compound C is weighed, only the mass m is known. But we can use the expression of the molar mass MM to calculate:

$$n = \frac{m}{MM}$$

So:

$$c_v = \frac{m}{V \cdot MM}$$

For the calculation of uncertainty Δc_v, we have to apply the rules of the Bureau International des Poids et Mesures (international office for measurements, BIPM, 2008) that expresses the uncertainty on a function of many variables (here c_v) as the square root of the sum of products: the square of the partial derivatives of the function relative to each variable, multiplied by the square of the uncertainty on each variable. Here we would have:

$$\Delta c_v = \sqrt{\left(\frac{1}{V \cdot MM}\right)^2 \cdot \Delta m^2 + \left(\frac{m}{V \cdot MM^2}\right)^2 \cdot \Delta MM^2 + \left(\frac{m}{V^2 \cdot MM}\right)^2 \cdot \Delta V^2}$$

2. We compare this result with the one that we calculate when a scale is used for all processes. For the mass fraction:

$$c_m = \frac{m}{M + m}$$

where M is the mass of the solvent (we could have calculated instead the quantity $c = \frac{m}{M}$).

Here, the uncertainty is:

$$\Delta c_m = \sqrt{\left(\frac{1}{M + m} - \frac{m}{(M + m)^2}\right)^2 \cdot \Delta m^2 + \left(\frac{m}{(M + m)^2}\right)^2 \cdot \Delta M^2}$$

Let us make this realistic: for the uncertainties on masses, we use the limit of the scale, and for the volume, we consider that the difference in levels (because of temperature driven dilatation) for a temperature difference of 10°C (between a room at 25°C, and a laboratory kept at 18°C), in a graduated flask.

How to compare these two methods? Of course, because c_v and c_m do not have the same units, you cannot compare them directly. On the other hands, you can compare $\frac{\Delta c_v}{c_v}$ and $\frac{\Delta c_m}{c_m}$. If you finish the calculation, you will have to use data about the thermal volume dilatation of the liquid (for ethanol, for example, it is 0.00109 /°C)(Engineering ToolBox, 2009):

$$v(T) = v_{T_0}(1 + \alpha \cdot (T - T_0))$$

Can you imagine the end of this calculation?

6.1.2 The Question

Previously, the distance between molecules in air was calculated, but what about this distance in water? Asking this question is helpful, because:

- it improves the representation ("theory") that we have of liquids (and remember the fifth culinary commandment: most food ingredients and foods are primarily composed of water);

- some students confuse this distance with the "length of the hydrogen bond," as if water molecules were as rigid as fixed as atoms in water molecules.

Hence the clear question: how far away are water molecules from each other in ordinary liquid water, in standard conditions?

6.1.3 Analysis of the Question

6.1.3.1 *The Data is Introduced*

No data is needed, because everyone knows that the mass of 1 L of water is 1 kg (OK, it depends on temperature, but we shall calculate here an order of magnitude).

Also, the Avogadro number is going to be used. For such a crude calculation, it can be chosen as 6×10^{23}, or more simply 10^{24} if order of magnitudes are used.

6.1.3.2 *Qualitative Model*

On the Internet, one can find models for water such as in Figure 6.1.

In such models, molecules are moving randomly in all directions with different velocities. Moreover the molecules are complex in geometry. For the calculation, we take a snapshot of water, which means that we consider fixed molecules, but we also decide to distribute the molecules on the centers of elementary cells of a square grid ("calculation on lattice") (Atkins, 1984) as a first step (Figure 6.2).

6.1.3.3 *Quantitative Model*

As before, we simply introduce symbols for representing what we see on the picture. Let it be:

- V the volume of liquid.

- ϱ the density of water.

- N the number of molecules.

- v the volume of an elementary cell.

- c the side of such a cell.

- N_A the Avogadro number.

6.1.4 Solving

6.1.4.1 *Looking for a Solving Strategy*

Assuming the regular distribution of molecules at the center of cubic cells, we see that the distance between two molecules is equal to the side of the cells.

How to determine the side of the cells? This can be determined from the volume of the cells.

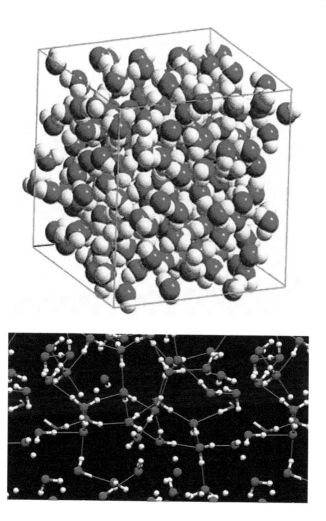

FIGURE 6.1　*(Top) Molecular modeling of water molecules in the liquid state at the temperature of 300 K. Of course, this picture is a snapshot of a dynamic system, and it is better to take sometime to look at videos such as https://www.youtube. com/watch?v=Zl74NCVbA5A. However, videos have to be interpreted because they are the results of calculations for only some picoseconds $(10^{-12}$ s). This means that complementary information about the velocities of water molecules is needed in order to describe the system correctly. In this picture, contrary to Figures 5.5 and 5.6, the background is not white, and frequently students asked about the "substance" between molecules answer "air" or "hydrogen bonds". This calls for a further discussion. At the bottom, another representation shows the hydrogen bonds.*

And there is a relationship between the volume of one cell and the total volume of the liquid: it is this volume divided by the number of molecules.

Now the question is to determine the number of molecules, and this can be done using the mass of the liquid and the density.

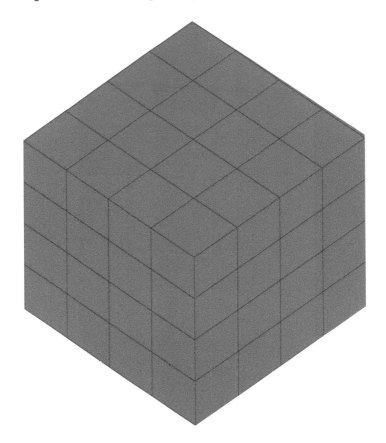

FIGURE 6.2 *Calculating on a lattice. This is obtained by applying the strategy of making a simpler description: instead of having a random dispersion, the molecules are regularly organized, and instead of random distances, the average distance is chosen. Each elementary cube has a water molecule in its center.*

6.1.4.2 Implementing the Strategy

The mass M of the volume V is given by the definition of density:

$$\varrho = \frac{M}{V}$$

It is:

$$M = \varrho V$$

In this mass, the number of moles is:

$$n = \frac{M}{MM} = \frac{\varrho V}{MM}$$

where MM is the molar mass.

And the number of molecules in the volume V is:

$$N = nN_A = \frac{\varrho V N_A}{MM}$$

where N_A is the Avogadro number.

The volume of one cell is the total volume divided by the number of molecules:

$$v = \frac{V}{N} = \frac{V}{\frac{\varrho V N_A}{MM}} = \frac{V\,MM}{\varrho V N_A} = \frac{MM}{\varrho N_A}$$

Thus, the distance between two water molecules is the cubic root of this volume:

$$c = \sqrt[3]{\frac{MM}{\varrho N_A}}$$

6.1.4.3 *Validation*

Here is the right place to note that many students have difficulty finding validations of their calculations, in particular because when a calculation is simple, there is not always two (or more) different ways to calculate the same value.

In such cases, order of magnitudes can be used to check the numerical application of the formal formula, as it is another way of doing.

For example, here, we would say that 1 kg has a volume of 10^{-3} m^3. And because the molar mass of water is 18 g, *i.e.* w 20 g, there are 50 moles of water in this volume, *i.e.*, $50 \times 6 \times 10^{23} = 3 \times 10^{24}$ molecules of water.

Then the volume for one molecule (in m^3) is 10^{-3} divided by 3×10^{24} which makes about 0.3×10^{-27}, hence the cubic root (in m): about 7×10^{-10}. This is about four times the length of a covalent C-C bond (0.15 nm, Chemistry-LibreTexts, 2020).

This new calculation is about the same as before, except that it is digital, instead of formal. Can we find something else? As said, looking for units is a possibility. We start from the formula that we got:

$$c = \sqrt[3]{\frac{MM}{\varrho N_A}}$$

Here, MM is a mass (kg). It is divided by a density (kg $\cdot$ m^{-3}) and because the Avogadro number is dimensionless, we find a value in m^3 under the cubic root. This means that the final result is in m, as awaited.

6.1.5 Expressing the Results

6.1.5.1 *The Formal Result Found is Written Down Again*

When studying simple questions, as the one discussed here, this step seems useless, but when problems are more complex, it is always a way to "clear the desk", and make the goal clearer.

$$c = \sqrt[3]{\frac{MM}{\varrho N_A}}$$

6.1.5.2 *Finding Digital Data*

Students know the molar mass of water, or they can find it by adding 16 + 1 +1 (g). Of course, it is useful to have a mnemonic for remembering the atomic masses of elements. Perhaps the easiest is to "say" simply "H he libeb cnofne namgalsip sclarkca sc tivchroman feconicuzn", so that you know the order for the first 30 elements, which is enough for many cases.

Molar masses are often given in g. Thus, in kg, the value would be 18×10^{-3}.

For the density of water, students generally know that 1 L of water contains 1 kg. They also know the Avogadro number.

6.1.5.3 Introduction of Data in the Formal Solution

$$subs\left(MM = 18e-3, \varrho = 1e3, N_A = 6e23, \sqrt[3]{\frac{MM}{\varrho N_A}} \right)$$

$$3.11 \times 10^{-10}$$

This result is given in SI units, *i.e.*, in m.

It can be observed that this is about twice the diameter of a water molecule, assuming that the order of magnitude of a covalent C-C bond is $1.5 \ 10^{-10}$ m. In reality, the length of the O-H bond in the water molecule is 0.99×10^{-10} m (see my comments in Cabane and Vuilleumier, 2005, and compare with Bieze *et al.*, 1993).

6.1.5.4 Discussion

In such a simple calculation, there is not much to explore, except when one uses tables of density as a function of temperature, and then the variations can be easily calculated.

For Table 6.1, we see that variations of density as a function of temperature are between 0.99984 and 0.95840.

Using the previous result, we can calculate the average distance between water molecules for these two extreme values:

$$subs\left(MM = 18e-3, \varrho = 0.99984e3, N_A = 6e23, \sqrt[3]{\frac{MM}{\varrho N_A}} \right)$$

$$3.11 \times 10^{-10}$$

$$subs\left(MM = 18e-3, \varrho = 0.95840e3, N_A = 6e23, \sqrt[3]{\frac{MM}{\varrho N_A}} \right)$$

$$3.15 \times 10^{-10}$$

It is observed that the variation is about 1%. This is not much, considering that we used orders of magnitudes for the calculations.

Now a good practice is to go online, in sites for scientific bibliography, and make a research with keywords such as "water"+"molecule"+"length of the OH bond", for

TABLE 6.1 The Variation of Some Properties of Water as a Function of Temperature in the Range 0–100°C.

t °C	Density (g/cm³)	Cp (J.g K)	Vapor Pressure (kPa)	Viscosity (μPa s)	Thermal Conductivity (mW/K.m)	Dielectric Constant	Surface Tension (mN/m)
0	0.99984	4.2176	0.6113	1793	561.0	87.90	75.64
10	0.99970	4.1921	1.2281	1307	580.0	83.96	74.23
20	0.99821	4.1818	2.3388	1002	598.4	80.20	72.75
30	0.99565	4.1784	4.2455	797.7	615.4	76.60	71.20
40	0.99222	4.1785	7.3814	653.2	630.5	73.17	69.60
50	0.98803	4.1806	12.344	547.0	643.5	69.88	67.94
60	0.98320	4.1843	19.932	465.5	654.3	66.73	66.24
70	0.97728	4.1895	31.176	404.0	663.1	63.73	64.47
80	0.97182	4.1963	47.373	354.4	670.0	60.86	62.67
90	0.96535	4.2050	70.117	314.5	675.3	58.12	60.82
100	0.9584	4.2159	101.325	281.8	679.1	55.51	59.81
Reference	Harr et al. (1984); Marsh (1987); Sengers and Watson (1986)	Marsh (1987)	Harr et al. (1984); Sengers and Watson (1986)	Sengers and Watson (1986)	Sengers and Watson (1986)	Archer and Wang (1990)	Vargaftik et al. (1983)

Note: All values (except vapor pressure) refer to a pressure of 100 kPa.

example. And one quickly finds documents (Huang *et al.*, 2013) providing the results of measurements: 0.311 nm. You see how interesting lattice calculation is!

6.1.6 Conclusions and Perspectives

A good exercise, after such calculations, is to use the result to revise the models (our initial picture). Of course, after having considered immobile molecules, it would be good to move toward molecular dynamics. Do you know the *Gromacs* package, for this (Gromacs, 2022)?

6.2 WHY WATER HAS NO COLOR AND WHIPPED EGG WHITES ARE WHITE? IS THIS DUE TO PROTEINS IN SOLUTION?

6.2.1 An Experiment to Understand the Question

6.2.1.1 *First, the General Idea*

Water is often depicted as blue, but in a glass, there is no color. Is really water blue?

1. Find a clear polycarbonate fluorescent lamp tube guard or other similarly sized clear plastic tube. Close one of its ends with a transparent film that you seal with a tape, so that you can pour water in the tube.

2. Put the tube vertically over a white sheet of paper, and take a picture with your mobile phone.

3. Add 2 cm of water, and take another picture.

4. Adding more water, take more pictures associated with various water heights, taking care that the light of the room does not change.

   You see: there is a color, in water.

6.2.1.2 *Questions of Methods*

The previous experiment is certainly an introduction but it lacks the rigor of the natural sciences of nature; as said previously, it is good to learn to avoid the use of adjectives and adverbs, and instead answer the question "how much?" The water is green? How much? Blue? How much?

Indeed, this question of the color of the water is important for deep sea exploration (NOAA, 2014), and we can improve the previous experiment by using free software such as *ImageJ* (NIH, 2020) to analyze these pictures.

One can either use the RGB (red, green, blue) system, or, when using a colorimeter, the L^*, a^*, b^* system (where L^* is for luminosity, a^* is an axis between red and green, and b^* an axis between blue and yellow).

6.2.1.3 *Where the Beauty Lies*

Now, it would be helpful to use the mobile phone again not only to take pictures of the various thicknesses of egg whites, but also of various thicknesses of foam.

It would spoil the calculation to give the result now, but let us add a hint: take pictures of a whipped egg white and also of the light used to illuminate it.

6.2.2 The Question

In food science and technology, students learn very quickly that egg white is made of 90% water, and 10% proteins (This vo Kientza, 2021a). Some conclude that proteins are responsible for the color of egg whites or even whipped egg whites. Here, we introduce calculations about this.

6.2.3 Analysis of the Question

6.2.3.1 The Data is Introduced

No data is given for this exercise.

6.2.3.2 Qualitative Model

If the question is about the visual appearance, the interaction of light with the material should be considered. Hence, the analysis should focus on these two aspects (first light, then the material).

In certain circumstances, light can be considered as a wave (McQuarrie and Simon, 1997). We know that waves of large wavelengths move almost without disturbance around small sized objects (like the swell around a pole, near the sea shore), but waves are reflected when their wavelength is short compared to the diameter of the objects that they meet. Here, we have to consider the size of visible light and to compare it to the size of water molecules or to the size of proteins dissolved in water for egg whites.

Of course, this description is not enough: for ions such as Cu^{2+} there is a color, for other reasons (Chemistry Stack Exchange, 2021).

Finally, air bubbles can be much bigger than molecules, as they can be seen with the unaided eye when whipping: at the beginning of the process, air bubbles are big. Then, when they are divided, their size becomes too small to be visible (remember this order of magnitude: the smallest visible object with unaided eye is about 0.1 mm, Carpenter *et al.*, 1845), but a microscope shows that they are visible with moderate magnification.

6.2.3.3 Quantitative Model

The wavelength of visible light is between 400 (violet) and 700 (red) nm (Feynman *et al.*, 1963).

Objects with a diameter of 0.1 mm can be seen with the naked eye at a distance of half a meter (Carpenter *et al.*, 1845).

Proteins are compounds whose molecules are made of more than one hundred residues of amino acids (indeed, the IUPAC observes that the molecular weight is greater than about 10000; for smaller oligomers, they would be called "peptides") (UPAC, 2019).

6.2.4 Solving

6.2.4.1 Looking for a Solving Strategy

We shall now compare various sizes.

6.2.4.2 Implementing the Strategy

The shortest wavelength in visible light (violet) is 400 nm. It is much bigger (about 4000 times) than the size of a water molecule, with chemical bonds, *i.e.*, 10^{-10} m (see Section 6.1.), so that the light is not perturbed: water appears transparent, even if, as observed, the color can change due the absorption of some electromagnetic radiations.

But how is it compared to the size of proteins? The molecules of proteins can be grossly divided into fibrillar and coils, but in egg white, most are coiled, or "globular proteins" (Sun *et al.*, 2020). Ovalbumin, for instance (45% of the protein content of egg white) has a molar mass of 45,000 g (Stein and Leslie, 1992).

Now, it is good to remember that the protein molecules are mainly made of residues of amino acids, for which the general formula is: $C(COOH)(NH_2)HR$, with the lateral group R = H for glycine, the simplest of all natural amino acids.

Here, a new little digression: you will notice that I have spoken of amino acid residues, and not of amino acids (but I gave the formula of an alpha amino acid). I did so because it is wrong to say that proteins are sequences of amino acids, just as it would be wrong to say that water contains dioxygen or dihydrogen: once dioxygen and dihydrogen have reacted, and water molecules have been formed, atoms have been rearranged, and some are even lost (during the formation of the peptide bond). In water molecules, there is no dioxygen, and no dihydrogen. And in protein molecules, even if one "recognizes" parts of amino acids, they are no more than residues. The same is true for triglycerides: they do not contain fatty acids, but only fatty acid residues.

Coming back to glycine, for example, the molar weight can be calculated:

$12 + 12 + 16 + 16 + 1 + 14 + 2 + 1 + 1 = 75$. Assuredly, part of it is lost during the making of the peptide bonds, on both side of the glycine residue), but the order of magnitude is 100 anyway. For other amino acids, the molar weight would be higher than that for glycine, but not by much.

Let us assume a mass of 150 for one amino acid residue. As the molar mass of ovalbumin is 45,000, this would mean about 300 residues for ovalbumin (indeed, the exact number is 385).

Now, in order to determine the size, we can imagine that this protein molecule is:

- either fully extended.

- or fully packed.

For the full extension, we can count three covalent bonds by amino acid residue, with an angle of 109.5° between bonds, because of the sp3 hybridization of orbitals on carbon atoms (McQuarrie and Simon, 1997). This would be about three times the length of the covalent bond (0.15 nm) multiplied by the sine of half the angle, *i.e.*, about 0.3 nm. And the total length would be about 90 nm, which is much shorter

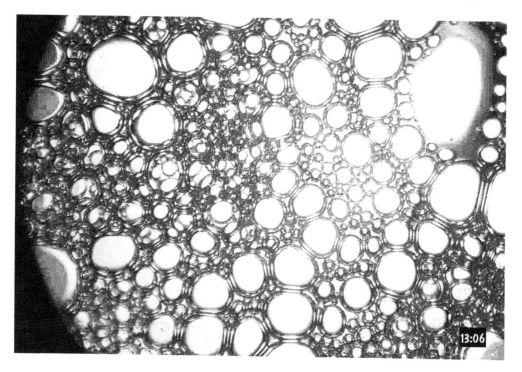

FIGURE 6.3 *On this microscopic picture of a whipped egg white, the largest bubbles have a diameter about 100 times larger than the smallest, but all show the reflection of the light of the lamp.*

(shortest than four times: remember, we said no adjectives and no adverbs) than the wavelength of light.

For the packed proteins, again a lattice calculation is helpful. Let us imagine that proteins are chains whose rigid segments fill up closely a small part of a cubic network (without overlapping). For each elementary cube, there are 12 segments, but each segment belongs to 4 cubes, so that one cube needs only 12/4, *i.e.*, 3 segments. And the number of elementary cubes corresponding to a fully coiled protein is $300/3 = 100$.

The size of a group of 100 elementary cells is $evalf(\sqrt[3]{100}) \approx 5$. This gives a total diameter of the coil about $5 \times 3 \times 0.15 = 2.25$ nm: this is much smaller than the smaller wavelength (violet).

In contrast, when egg whites are whipped, the bubbles are initially between 1 mm and 3 cm, and then reflections of the ambient light, be they lamps on the ceiling or sunlight through windows, can be observed on each bubble. Whipping introduces more and more bubbles (as well as reflections), as it divides bubbles repeatedly. However, eventually the size of air bubbles is barely visible with the naked eye, and the optical microscope still shows reflections on any bubble (Figure 6.3).

In Figure 6.3, the largest bubbles have a diameter of about 0.1 mm (10^{-4} m), and the smallest of about 10^{-6} m. The latter remains larger than the wavelength: visible light reflects on bubbles walls so that if this light is blue, the whipped egg white appears blue, red if the light is red, and white if the light is white.

6.2.4.3 Validation

For validation, the students are encouraged to do a bibliographical survey on light absorption.

6.2.5 Conclusions and Perspectives

This exercise gives an order of magnitude of various important objects for food science and technology.

Other sizes can be calculated in this way: food science and food technology are full of objects of interest, from the supramolecular association of phenolics with metals (Fulcrand *et al.*, 2006) to the macroscopic size of food in the plate.

Indeed, such calculations are useful for making more precise models of colloidal systems, such as the many that can be made about food.

6.3 IS IT POSSIBLE TO DRY TOMATOES IN AN OVEN AT 105°C IF THE DOOR OF THE OVEN IS CLOSED?

6.3.1 An Experiment to Understand the Question

6.3.1.1 First, the General Idea

Sun-dried tomatoes are delicious, but expensive when you do not make them by yourself. If you live in a country where the sun is not hot enough to make them, you can use an oven. Here is the recipe:

1. In a pan, put water to a boil.

2. Then, dip tomatoes in this boiling water for less than 20 s.

3. Peel the tomatoes (after boiling, the skin is easy to peel.

4. Cut the tomatoes in halves, and press them to eliminate their liquid content, and the seeds.

5. Put the halves on a baking plate, cut face up. Sprinkel with sugar, salt, finely cut garlic, thyme and rosemary, plus olive oil.

6. Heat the tomatoes in the oven at 105°C, but open the door regularly in order to allow the vapor escape.

6.3.1.2 Questions of Methods

About peeling tomatoes, here is a possibility of complementary research for classes: how long should they be boiled to make them easy to peel? Above, we indicated a time less than 20 s (based on an experimental seminar that I organized, with many tomato varieties and conditions), but please repeat the experiments in order to validate ours.

Of course, you cannot compare too many tomatoes and working with a class is better because you can make statistics. One group could test 10 tomatoes (if possible from different qualities and batches) boiled for 5 s, another group could test 10 tomatoes boiled for 10 s, and so on by 5 s. Finally, with this collaborative research, the question would be experimentally solved.

6.3.1.3 *Where the Beauty Lies*

In the kitchen, the main point about "beauty" is not the one that you see on your plate. Just as in music beauty is beauty to hear, in cooking it is "beauty to eat" (it has to be "good"). And dried tomatoes are better when correctly seasoned, and also when they are used smartly.

Let us give a recipe using the dried tomatoes that we made before. For example, dice them and add them to a green salad, along with a tapenade, that will be obtained by grinding about 150 g of black olives with the garlic slices from the dried tomatoes, the juice of one lemon, a drop of an anise brandy (Pastis, for example), about 50 g of stale bread, plus a large glass of olive oil (to be added slowly during grinding, in order to get an emulsion).

But we do not forget that this book is not only about cooking, but also about sciences of nature. Here, we can discuss usefully the possible results of the experiment that would be done according to the previous paragraph. Certainly, there will be some differences in the various batches, and some statistics will be needed. We shall not make here a whole course in this, but we take the opportunity to indicate how variations can be important for small food samples. This is useful when one uses a precision scale, for which factors of variation are important: which value should we use?

Of course, the answer is: where there is no obvious mistake, we should use the average value of many successive measurements and keep along an estimation of the dispersion, *i.e.*, the standard deviation for the measurements. Students learn that 70% of a population lies within one standard deviation around the average, and 95% within two standard deviations (Reinhart, 2022), but they often are surprised by the possibilities of large deviations.

Let us explore this digitally by creating first a normal (or Gaussian) population (distribution as e^{-x^2}) with an average of 100, and a standard deviation of 1: in this way, we do not need a precision scale. Imagine now that we take three samples from this population at random (the equivalent of weighing three times the same object on the same balance) and that we determine the average of the three values, as well as their standard deviation. Figure 6.4 shows the distribution of the calculated standard deviations for 200 successive determinations of this kind.

You see how the standard deviations are dispersed!

Now, if one picks at random samples with 6 elements in each group, the average value does not change much, but of course the dispersion of the standard deviation values is narrower (Figure 6.5).

And with even bigger sample, the distribution of standard deviations is even more narrow (Figure 6.6).

This kind of digital experiment is helpful because it shows that, for small samples, one would be well advised to consider that the standard deviation of the sample is only an order of magnitude of the standard deviation for the population. In particular, this explains why is is nonsense to consider too many digits for standard deviations.

6.3.2 The Question

Let us now go back to the question of drying tomatoes in an oven. Is it possible to dry the tomatoes in an oven if the door remains closed?

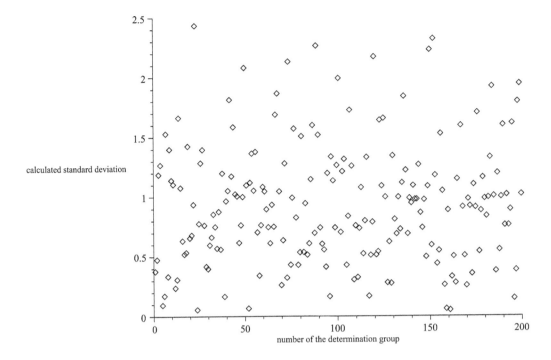

FIGURE 6.4 *Standard deviations for 200 samples of 3 elements only (in the lab, when one uses a precision scale, it is a frequent practise that three measurements are performed).*

6.3.3 Analysis of the Question

6.3.3.1 The Data is Introduced

The temperature of the oven is given: 105°C. Of course, tomatoes at the sun would not reach such a high temperature, but it does the job well, as you will see if you make really the experiment.

6.3.3.2 Qualitative Model

One could study this question using the phase diagram of water, considering the vapor pressure at ambient pressure and at the temperature given for the oven. However, if the oven were closed, the pressure would increase in a very unrealistic way.

Other possibilities exist for solving this question. In particular, we could assume that the pressure could push the door open, so that the vapor escapes until the tomatoes are entirely dry: assuredly, this is not wise from a culinary point of view, but it provides another possible solution to our questions, and this is the most important lesson of this book: do the solution that you are able to do.

In passing, let us observe that for ovens, gas systems are not the best, because gas burning creates water vapor in the oven; they have vent tubes that it is important to open. And many electric ovens have safety systems that stop heating when the door is opened; here, you will have to open the door regularly for some seconds.

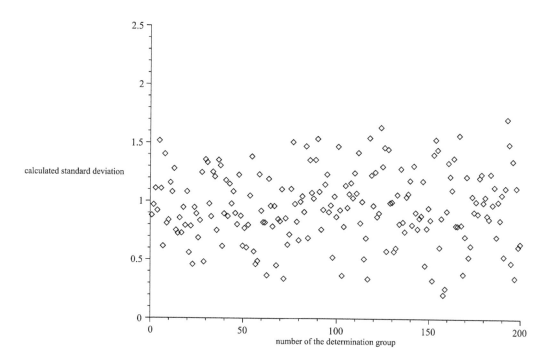

FIGURE 6.5 *Standard deviations for 200 samples of 6 elements.*

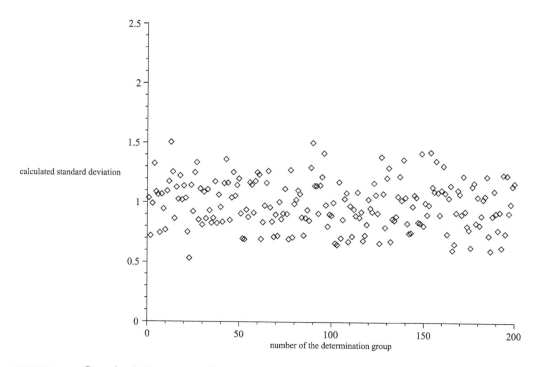

FIGURE 6.6 *Standard deviations for 200 samples of 15 elements.*

FIGURE 6.7 *An oven with drying tomato. In such pictures, dimensions do not need to be respected. Only avoiding the possibility of making a wrong symbolic description is important.*

Finally, let us study the following question: is there enough vapor pressure in the oven so that the door is pushed, allowing the escape of steam during cooking? One can observe that this question is the same as: does water evaporates from a pan on which a lid is placed?

Or, in lab activities, should we cover with a watch glass a beaker in which a sample should dry (see the AOAC method: the sample should be put in the oven at 105°C)(AOAC, 2000)? In this case, the answer is yes, because one has to avoid deposition of foreign material in the beaker (we had flies, once!). A picture is to be made (Figure 6.7).

6.3.3.3 Quantitative Model

We consider a mass m of tomatoes.
The water content (proportion of the fresh mass) of tomatoes is p.
The temperature in the oven is T.
The area of the door of the oven is A.
The volume of the oven is V.

6.3.4 Solving

6.3.4.1 Looking for a Solving Strategy

It is proposed to consider the forces needed in order to open the door. This force comes from the pressure increase due to water evaporation.

Of course, the more tomatoes from which water is evaporated, the higher the pressure.

6.3.4.2 Implementing the Strategy

The initial pressure is the atmospheric pressure acting on both sides of the oven's door. Now, assuming that there is no exchange between the oven and the outside, the evaporation of water adds to the inner pressure.

If a mass μ of water is evaporated from the tomatoes, this correspond to a number n of moles equal to μ/M, where M is the molar mass of water.

And steam is a gas. If we assumed that it behaves as an ideal gas:

$$PV = nRT$$

the pressure is:

$$P = \frac{nRT}{V} = \frac{\left(\frac{\mu}{M}\right) \cdot RT}{V}$$

This pressure corresponds to a force F that can push the door (area A):

$$P = \frac{F}{A}$$

Thus:

$$\frac{\mu RT}{MV} = \frac{F}{A}$$

The force is:

$$F = \frac{\mu RT A}{MV}$$

A force (in newtons, N) is not something that one can grasp easily, and a mass (imagine a pulley) is probably more understandable. This mass ν can be found using its weight:

$$F = \frac{\mu RT A}{MV} = \nu g$$

$$\nu = \frac{\mu RT A}{MVg}$$

6.3.4.3 Validation

Here, it is proposed that the result is simply corroborated using numerical applications.

6.3.5 Expressing the Results

6.3.5.1 The Formal Result Found is Written Down Again

$$\nu = \frac{\mu RT A}{MVg}$$

6.3.5.2 Finding Digital Data

Let us imagine that we dry 2 kg of tomatoes. The water content of tomatoes is about 95% (Ciqual, 2020).

The constant g is known to be equal to about $9.8 \text{ m} \cdot \text{s}^{-2}$.

The oven is assumed to be a cube of 0.5 m for its side.

The molar mass of water is 18 g, to be expressed in kg: 18×10^{-3}.

The temperature is given as initial data: 105°C, or 378 K.

The constant of ideal gases is taken equal to 8.32 J/mol $\cdot$ kg.

6.3.5.3 Introduction of Data in the Formal Solution

Below, the result is given in SI units, *i.e.*, kg

$$subs\left(\mu = 0.95 \cdot 2, g = 9.8, V = 0.5^3, M = 18e - 3, A = 0.5 \cdot 0.5,\right.$$

$$\left. R = 8.32, T = 378, \frac{\mu R T A}{M V g}\right)$$

$$6.77 \times 10^4$$

About 70 tons! This means that evaporating the full content of water would be enough to open the door, but indeed ovens are generally full of openings through which water can escape.

6.3.5.4 Discussion

As said previously, this kind of exercise can be dealt with in a number of different ways, and the more the better. For example, one could use a thermodynamic approach, considering the equilibrium liquid-gas for water. When no data is given, one could simply interpolate between the vapor pressure of water at 20°C and at 100°C: one could even assume (this is obviously not true) that the saturating vapor pressure at 20°C is 0, and that it is (this is true) 10^5 Pa at 100°C. Then, one would consider the volume of the oven and write the ideal gas model for finding the number of moles of water vapor. By subtraction, one could see whether the tomatoes can be dried.

For discussing this question, one could also calculate the cost of drying tomatoes: this is given in the Annex below.

6.3.6 Conclusions and Perspectives

6.3.6.1 Conclusions

In the "Conclusion" sections of the "backbone for calculation", one is invited to discuss the result. And here, you can come back to the main question, about the possibility of drying tomatoes in a closed oven: yes, you can.

But let us observe that you considered the door as a lid, instead of a door that turns around a hinge. If you want to review some mechanics, do it, assuming a return force for the door (imagine a spring in the system).

6.3.6.2 Perspectives

With this question, we wanted to show that simple culinary questions can be investigated, and at each step, there is a possibility to delve deeper. The issue is simply to look for knowledge that one has and then implement it.

6.3.7 Annex: How Much Does it Cost to Evaporate Water?

6.3.7.1 *The Question*

Assuming that we paid for tomatoes, drying them under the sun does not cost any-thing (except the immobilization of assets), but how much does it cost, at home, using an oven? This question is important to refine ideas with orders of magnitudes.

6.3.7.2 *Analysis of the Question*

6.3.7.2.1 The Data is Introduced

The price of energy is easily be found online. For example 0.19 euros per kWh (Eurostat, 2020).

And 1 kWh = 1000 × 3600 J

So that the price per S.I. unit of energy (euro per joule) is:

$$pp = \frac{0.19}{1000 \cdot 3600}$$
$$pp = 4.17 \times 10^{-8}$$

6.3.7.2.2 Qualitative Model

In Figure 6.8, a dynamic process can be envisioned.

We shall calculate the full drying.

6.3.7.2.3 Quantitative Model

The initial temperature is T_1.
The initial mass is M_1.
The final temperature is T_2.
The final mass is M_2.
The initial proportion of water in tomatoes is p.

6.3.7.3 *Solving*

6.3.7.3.1 Looking for a Solving Strategy

We need to calculate the amount of energy needed to evaporate the water of tomatoes. Tomatoes have to be heated to a temperature of 100°C, then water transforms from the liquid state to the vapor state. This involves two steps:

1. heating tomatoes to the boiling point of water;

2. evaporating water.

6.3.7.3.2 Implementing the Strategy

For the first step, the energy needed is:

$$E_1 = mc_p \Delta T$$

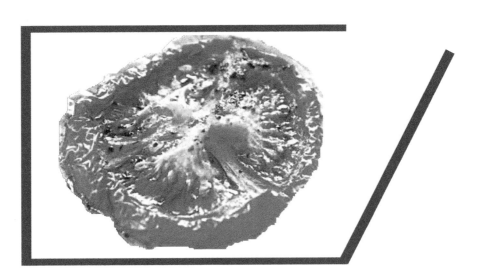

FIGURE 6.8 *Drying tomatoes means changing the content in water.*

where m is the mass (tomatoes) to be heated, c_p the heat capacity of tomatoes (assumed to be the same of water because of the high water content in tomatoes) and ΔT the temperature difference between room temperature and 100°C.

When the boiling point is reached, the energy used for evaporating water is proportional to the mass, but in tables, one often finds it per number of moles:

$$E_2 = nL,$$

where L is the latent heat of vaporization (in $J \cdot mol^{-1}$).

Of course, we have:

$$n = \frac{m}{M}$$

where M is the molar mass of water.

So that, if pp is the price for one unit of energy, the price Pr for drying a mass m of tomatoes is:

$$Pr = pp(E_1 + E_2) = pp \left(mc_p \Delta T + \frac{m}{M} L \right)$$

6.3.7.4 Expressing the Results

6.3.7.4.1 The Formal Result Found is Written Down Again

$$Pr = pp \left(m \, c_p \Delta T + \frac{m}{M} L \right)$$

$$Pr = pp \left(mc_p \Delta T + \frac{mL}{M} \right)$$

6.3.7.4.2 Finding Digital Data

The latent heat can be found in tables of constants, but it is probably a good exercise to try to guess it, assuming that 2 hydrogen bonds per molecule (energy e) have to be broken during evaporation. Thus, for one mole:

$$L = 2N_A e$$

From Suresh and Naik (2000), we see:

$$e = 23.3 \text{ kJ} \cdot \text{mol}^{-1}.$$

Then:

$$L = 23.3 \cdot 1000 \cdot 2$$
$$4.66 \times 10^4$$

This value (in $J.mol^{-1}$) can be compared to the latent heat of water vaporization (Garai, 2009):

$$40.68 \cdot 1000$$

$$4.07 \times 10^4$$

It is reasonably close (13%).

6.3.7.4.3 Introduction of Data in the Formal Solution

$$evalf\left(subs\left(pp = 4.17\mathrm{e}-8, m = 2, c_p = 4181, \Delta T = 80, L = 40680, M = 18\mathrm{e}-3, pp \right.\right.$$
$$\left.\left. \cdot\left(m\,c_p\,\Delta T + \frac{m}{M}\,L\right)\right)\right)$$

$$2.16 \times 10^{-1}$$

And this result is in euros.

6.3.7.5 Conclusions and Perspectives

It would be useful for students try to evaluate their results, comparing them to real prices.

When the tomato season is on, the price for 1 kg of fresh tomatoes can be as low as about 1 euro.

On the other hand, the price for sun-dried tomatoes is about 3 euros per kilogram. This means that the price of energy is not the main factor for selling dried tomatoes.

6.4 CAN THE TEMPERATURE REACH 130°C IN A PAN WHERE WATER IS BOILING?

Here, let us do as in the section about flaming oil: we immediately give the answer: no!

No, in an ordinary pan, in ambient conditions, the boiling temperature of water is very close to 100°C, and this particular temperature is even used for calibrating thermometers and other devices for temperature measurement.

But we remember that the goal of this book is to invite you to make experiments and calculations. First, we need to know why this question was proposed and secondly, we need to experiment and calculate.

6.4.1 An Experiment to Understand the Question

6.4.1.1 First, the General Idea

The question, here, is about the relationship between pressure and temperature: one knows that water boils at less than 100°C in high mountains, and that the boiling temperature is increased when pressure is applied (pressure cookers). But for ordinary pans with simple lids put on them, the question studied seems nonsense. Why consider it?

Primarily because this 130°C story is from a starred chef! Could one of these remarkable artists be wrong? Are we allowed to think that famous individuals that have so many fans can distribute inaccurate information publicly?

The easiest answer is to make the experiment, and validate it by calculation (let us always be cautious about knowledge: check and validate):

1. in a pan, put some water;

2. measure the temperature;

3. put to a boil;

4. measure the temperature again.

And you will see that the boiling temperature is close to 100°C.
Now, compare with this second experiment:

1. in a pan, put some water;

2. measure the temperature;

3. cover with a lid;

4. put to a boil;

5. take the lid out and measure the temperature of water rapidly.

You will see that the boiling temperature is again close to 100°C.
Because such an experiment is not very entertaining and because it is sometimes said, in kitchens, that salt can change the boiling temperature, let us now make a different experiment:

1. in a glass, add 200 g of water;

2. measure the temperature;

3. add 200 g of salt and stir in order to reach the maximum dissolution;

4. measure the temperature again;

5. put this salted water in a pan, and bring to a boil;

6. measure the temperature;

7. cover with a lid and observe;

8. rapidly, take the lid off and measure the temperature;

9. put the lid back on the pan, with a heavy weight on it, and observe.

6.4.1.2 Questions of Methods

With the last experiments, there are many interesting questions:

1. When salt is added to water, the temperature of water (becoming a solution) decreases: do you know why? By the way, with saturation, as in the previous paragraph, one can observe the maximum decrease (for salt). Would you like to compare with the addition of sucrose (table sugar) in water?

2. With a lot of salt in water, the boiling point of water is increased: do you know why? And which temperature would one reach when heating?

Here, it is good to observe that in the past, there were only two well known temperatures in kitchens: 0°C, when ice is melting in liquid water, and 100°C when water boils. And it is always good to remember that measurement tools, such as thermometers, have to be controlled: personally, I have seen a professional pastry thermometer that displayed 90°C in boiling water, at low altitude! For sure, the variable atmospheric pressure can change the boiling point, but how much? If you calculate it, you will find that we are right to keep sound ideas such as "water boils at 100°C".

As water boils ordinarily at 100°C, it explains why it is a good practice to boil eggs in already boiling water, compared to putting eggs in cold water and heating it: the eggs spend more time over 60°C in the second case (they begin coagulating at this temperature), and the result is less precise. On the contrary, with a fixed 100°C temperature of cooking, the only parameter that can change is the size of eggs.

6.4.1.3 Where the Beauty Lies

Here we discuss two ideas. First, one is invited to make the complementary experiment of boiling a test tube in which 1 g of water was put, with a balloon closing the tube on top: when all water evaporates, one can observe the expansion of the balloon, with a final volume of about 1 L, if the balloon is very elastic. This is an important order of magnitude to keep in mind, and it corresponds to what we have to remember: 1 mole of water (18 g) generates 30.5 L of vapor at 100°C; this means that about 1 g corresponds to 1.5 L. Oh, by the way: could you calculate it from one of the 14 basic formulas?

Now, the last experiments can have consequences for testing what seaside populations sometimes say: sea water would reach boiling slower than non-salted water. Indeed, the experiment gives information for interpreting this saying: if the initial temperature with salt is lower than room temperature, and if the final temperature (boiling) is higher than 100°C, this would explain how the boiling time could be longer.

Except that one should define carefully the initial and final states! If the goal is to boil water that is finally salted, then the two effects of lowering and increasing the temperature occur in both cases. And indeed the experiment shows that the uncertainties are so great that no measurable difference can be determined.

Let us add that cooks often say that salt accelerates the boiling process because it is true that when salt is added to boiling water, more bubbles appear suddenly—but this would occur also with sand because vapor bubbles nucleate on impurities—and this does not change the boiling temperature.

6.4.2 The Question

We now repeat the question: a famous starred chef wrote in one of his books that in a pan full of water with a lid simply put on top of it, the temperature can reach

FIGURE 6.9 *Under the lid (black) on a pan with water (blue), part of the water is liquid, and part of the water is in the vapor phase.*

130°C. We are not giving the reference because the chef is alive, and it will be shown that what he wrote is nonsense.

Let us check what he wrote.

6.4.3 Analysis of the Question

6.4.3.1 The Data is Introduced

Only one data point is given: the temperature of 130°C, *i.e.*, 273 +130 = 403 K.

6.4.3.2 Qualitative Model

For the qualitative model, we said already that the first thing to do is to make a simple picture (Figure 6.9): the goal is not to make beautiful pictures, but to draw a meaningful one.

Hopefully, it is obvious for all science and technology students that the boiling temperature is $T = 100$°C at sea level, for the standard atmospheric pressure (101325 Pa), and that a temperature of 130°C can be reached only when the pressure is increased.

Pressure cookers are often equipped with a temperature measurement system, and always with a security valve, showing that the pressure is high. However, the chef whose writings we discuss was dealing with ordinary pans, not a pressure cooker.

6.4.3.3   Quantitative Model

Let us consider a pan of volume V, containing a volume v of water. The mass of the lid will be M, with area A.

6.4.4   Solving

6.4.4.1   Looking for a Solving Strategy

Starting from water and air, the steam formed through water vaporization contributes to increasing the pressure of the gas over liquid water (assuming that it does not escape first). The vapor pressure of water adds to the pressure of air, and this can lift up a lid when the total pressure is more than the one applied by the lid (its mass makes a force over the surface of the pan).

6.4.4.2   Implementing the Strategy

If the mass of the lid is M, the force on the gas is the weight W:

$$W = M \cdot g$$

From this force and the area A of the lid, we can determine the pressure exerted by the lid:

$$P = \frac{W}{A} = \frac{M \cdot g}{A}$$

Now, the pressure of the gas is the sum of all partial pressures, *i.e.*, 1 atm for the air, plus the vapor pressure due to water vaporization.

The question is now to find this vapor pressure, due to vaporization of water. Here, there is a sub-strategy to implement, because the pressure of water vapor depends on temperature, but the temperature (the boiling one) itself depends on pressure. We can envision various possibilities:

(1) assuming that the temperature is not much more than 100°C even with some pressure added to the atmospheric one, we calculate the mass of water that has to evaporate in order to lift the lid; for sure, this means that we shall not consider the 130°C given by the chef as a real possibility, and it will overestimate the pressure;

(2) assume that the temperature reaches 130°C, and calculate the mass of the lid that the pressure would lift up;

(3) use a phenomenological model, such as the Duperray relationship, between pressure and temperature, in order to be able to find all parameters (see below, and also McQuarrie and Simon,1997);

(4) use molecular modeling, in order to explore the question (Factorovich *et al.*, 2014).

Let us do that now.

(1) Assuming the temperature is 100°C, the ideal gas model for water vapor in the pan (over liquid water) is:

$$P \cdot V = nRT$$

Here, the pressure P can increase until the lid is lifted; at this point, the pressure of the gas is equal to the pressure exerted by the lift:

$$\frac{M \cdot g}{A} \cdot V = nRT$$

We determine the number of moles of water being evaporated to make this pressure:

$$n = \frac{MgV}{ART}$$

And finally, we use the molar mass of water MM to calculate the mass of water that evaporated:

$$m = nMM = \frac{MgV\,MM}{ART}$$

(2) If the temperature is assumed to be 130°C, we can calculate the pressure of vapor:

$$P = \frac{nRT}{V}$$

And this pressure is equal to the one of the lid:

$$P = \frac{M \cdot g}{A}$$

This means that we have:

$$\frac{nRT}{V} = \frac{Mg}{A}$$

Here, we can determine the mass of the lid assuming that different quantities of water evaporated:

$$M = \frac{nRTA}{Vg}$$

(3) The Duperray model states that

$$P = P^\circ \cdot \left(\frac{\theta}{100}\right)^4$$

where P° is the initial pressure (1 atm) and θ the temperature in °C; P is in atm. Using P, one could calculate the mass of lid that would hardly resist the pressure.

(4) Here is a large field for modern research, as shown in Chaplin (2020).

6.4.5 Expressing the Results

6.4.5.1 The Formal Result Found is Written Down Again

We limit this to the first three cases:

1. $m = \frac{MgV\,MM}{ART}$

2. $M = \frac{nRTA}{Vg}$

3. $\theta = 100 \cdot \left(\frac{\left(\frac{Mg}{A} \right)}{P\circ} \right)^{\frac{1}{4}}$

6.4.5.2 Finding Digital Data

All values in this result were previously used many times.

6.4.5.3 Introduction of Data in the Formal Solution

Let us use a pan of volume of $V = 5$ L, radius $r = 15$ cm, containing the 1 L of water.

The area of the lid would be:

$$A = \pi \cdot 0.15^2$$

We assume a mass of the lid $M = 0.1$ kg.

1. The mass of evaporated water can be determined:

$$subs\Bigg(M = 0.1, g = 9.8, A = \pi \cdot 0.15^2, V = 5e - 3, R = 8.32, T = 373,$$

$$MM = 18, m = \frac{MgV\,MM}{ART} \Bigg)$$

$$m = 4.02 \times 10^{-4}$$

Here the mass is in g because we used the molar mass value in g.

2. For this second calculation, we decided to determine the result for various masses of water evaporated:

- for 1.8 g:

$$subs\Bigg(g = 9.8, A = \pi \cdot 0.15^2, V = 5e - 3, R = 8.32, T = 373,$$

$$MM = 18, n = 0.1, M = \frac{nRTA}{Vg} \Bigg)$$

$$M = 4.48 \times 10^2$$

- for 18 g:

$$subs\left(g = 9.8, A = \pi \cdot 0.15^2, V = 5e-3, R = 8.32, T = 373,\right.$$

$$\left. MM = 18, n = 1, M = \frac{nRTA}{Vg}\right)$$

$$M = 4.48 \times 10^3$$

- for 180 g:

$$subs\left(g = 9.8, A = \pi \cdot 0.15^2, V = 5e-3, R = 8.32, T = 373,\right.$$

$$\left. MM = 18, n = 10, M = \frac{nRTA}{Vg}\right)$$

$$M = 4.48 \times 10^4$$

3. Here we calculate the pressure assuming a temperature of 130°C.

$$solve\left(P = \left(\frac{130}{100}\right)^4, P\right)$$

$$2.86 \times 10^0$$

This pressure is in atm.

We can now determine the mass:

$$subs\left(P = 2.86 \cdot 1e5, g = 9.8, A = \pi \cdot 0.15^2, M = \frac{PA}{g}\right)$$

$$M = 2.06 \times 10^3$$

A mass of more than 2 tons!

6.4.5.4 Discussion

Clearly, all calculations refutes the chef: in an ordinary pan, the temperature would never boil at 130°C.

6.4.6 Conclusions and Perspectives

With this exercise, we want to show that we have always to resist the "authority argument", in particular from chefs, even if they have Michelin stars. Indeed, chefs are more artists than technicians, and not all of them are competent in physical chemistry (yes : ;-)). Culinary books, as well, are full of mistakes and false interpretations (This, 2010).

The most common mistakes are about what they call "thermal shock", "Maillard reactions", "oxidation", "albumin", "chlorophyll" (This vo Kientza, 2021b). For example, it is often said that "Maillard reactions are responsible for meat browning during cooking", but such a sentence is wrong: first, the "Maillard reactions" should be called glycation reactions (Sharon, 1985) and they are only part of the story because many other chemical processes take place (This, 2016). About "albumin", this is an old chemical term (This, 2009), and today we would speak of "proteins", and "albumins" when referring to a category of globular proteins. The same holds with chlorophyll: this name is an old chemical name, but today chemists speak of chlorophylls (This, 2017). However, with elementary scientific knowledge and calculation skills, students in science and technology can have a critical view on what is written in culinary books. They can exert their judgment in order to become advanced scientists or technologists.

6.5 BY HOW MUCH IS THE ACIDITY OF WINE INCREASED THROUGH OXIDATION OF THE SULFUR DIOXIDE (SO_2) THAT WAS ADDED BEFORE CORKING?

6.5.1 An Experiment to Understand the Question

6.5.1.1 First, the General Idea

For this experiment, sulfur is needed: one can find it in drugstores, as a yellow powder, with interesting properties. Of course, before using it, one should look for the safety information (ECHA, 2020), and take safety precautions.

When this is done:

1. in the open air or under a hood, put about 1 g of sulfur in a test tube;

2. while you hold the tube with a non-conductive material (such as a wooden clip), heat the bottom of the test tube;

3. you will observe that the initially yellow powder turns red and liquid;

4. pour the content in cold water, and observe the consistency of the brown solid in water.

With this experiment, you will be able to observe an "allotropic change": sulfur becomes rubbery because it was changed from a solid with formula S into a another "allotropic" form, of formula S_8.

Then, let us go on with sulfur burning in air:

1. prepare a large beaker and a glass watch that will be able to cover it entirely;

2. with a lighter, heat some sulfur powder in a small cup until you observe a blue flame (do not smell it, because this is dangerous sulfur dioxide SO_2);

3. when sulfur is burning, put the cup in the beaker and cover with the glass watch;

4. when the flame goes out, add some drops of a dilute solution of potassium permanganate $KMnO_4$ in the beaker;

5. cover again, and observe: after some seconds, the purple solution of potassium permanganate is reduced into colorless manganese Mn^{2+}.

For this second experiment, the sulfur dioxide produced by burning sulfur in air (thanks to dioxygen) accumulated at the bottom of the beaker, because its density is more than the one of air (remember the formula $d = \frac{M}{29}$, where M is the molar mass of the compound making up the gas). Here, the density of sulfur dioxide is:

$$evalf\left(\frac{32 + 16 + 16}{29}\right)$$

2.2068965520

When reacting with potassium permanganate, sulfur dioxide is oxidized into sulfuric acid H_2SO_4, while the permanganate ion $MnO_4{}^-$ is reduced into colorless manganese Mn^{2+}. Can you equilibrate the chemical reaction?

6.5.1.2 Questions of Methods

For wines and food, the question of using sulfides as preservative is much discussed. Sulfur has been used as a powder deposited on cotton, that is burnt in wine barrels, in order to sterilize them, block oxidation reactions or stabilize wine (OIV, 2021). It is also used in the bee industry, to fight some bee diseases, for example (USDA, 2008).

But manipulating powders is not very easy, and can create hazards, such as explosions due to electrostatics or inhalation of suspended $K_2S_2O_5$ particles (SCAT, 2017).

For the wine or beer industry, potassium metabisulfite is used instead: it can dissociate in an acidic environment (remember the Section 5.4 in chapter 5: wine has often a pH lower than 4) to produce sulfur dioxide (EFSA, 2008a).

6.5.1.3 Where the Beauty Lies

There are many traditional experiments in chemistry using sulfur dioxide, but an interesting one is about bleaching the petals of roses. This can be done in the following

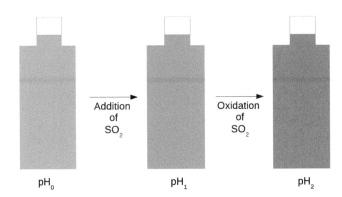

FIGURE 6.10 *Wine in a bottle. Making this simple scheme can avoid wrongly interpreting the question.*

way: sulfur is burnt in a test tube, and when it is burning, a red petal is put at the entrance of the tube. This petal turns white in some seconds. Then, the petal is washed in water, and stored in a dark box: after some time, it is red again.

All this leads to the remarkable story of the color of plants due to carotenoids and phenols (Eugster, 1991; Quideau, 2022), keeping in mind that red wines owe their color to these compounds. The change of color with time certainly occurs in a reducing environment, which explains why the question discussed in this chapter is so important.

6.5.2 The Question

It is a fact of observation that repeating clearly a question being studied and focusing on words is important for solving exercises. Indeed, during the preparation of this book, it was observed that, for reasons that will not be discussed here, some students or colleagues who were asked the question of this paragraph, calculated the acidification of water by sulfur dioxide (from pH = 7) instead of answering the question about the acidification by the oxidation of sulfur dioxide already added to wine, as asked in the title of the exercise.

6.5.3 Analysis of the Question

6.5.3.1 The Data is Introduced

No data is given, because students nowadays have broad access to information. Of course, the places to look can be the topic of some useful discussion. In particular, scientific sources found using *Google Scholar,* or *PubChem* for chemistry, or other scientific databases and engines are probably more acceptable than simply googling (even if this is a good starting point). In the past, *Wikipedia* was initially weak but with time, it has become more and more interesting.

Remember that information should be used only if it is justified by a "good" reference (official, science publications, etc.), or by a validated experimental method.

And concerning Internet sites, the date of last access has to be given (even better: the information is given in a standardized way YYYY-MM-DD).

6.5.3.2 Qualitative Model

When considering the particular question of this chapter, one can see that two steps can be considered (Figure 6.10):

1. dissolving sulfur dioxide in wine (for preservation) during wine making;

2. oxidation of sulfur dioxide when the bottle of wine is opened.

6.5.3.3 Quantitative Model

On the picture, we see that we have to introduce:

- the volume of wine V.

- the mass m of sulfur dioxide added to wine.

6.5.4 Solving

6.5.4.1 Looking for a Solving Strategy

Here, even if we know that calculating the pH of wine and its modifications is tricky, we will begin simply by making the initial calculation of the pH of water (pH $= 7$) to which sulfur dioxide is added. This looks like what we did in Chapter 5 (paragraph 5.4.); however, here, we will not simply apply a known formula. Instead, we shall make the entire calculation.

Then, we shall consider the oxidation of sulfur dioxide, producing sulfuric acid and changing the pH. The reason for this is that we want to demonstrate how pH calculations are easy using formal software such as *Maple* or another software of the same kind: using them should change the way pH calculations are taught in the university.

The real phenomenon will be considered in the Discussion section.

6.5.4.2 Implementing the Strategy

In the past, in order to calculate the pH of a solution, the old educational method was to write the various chemical equations of chemical phenomena occurring, express them in quantitative terms, and make assumptions in order to be able to solve the system of equations that was obtained. Often, students succeeded in the two first steps, but failed in the last one so that they relied on memorizing the final formulas (complaining that chemistry was uninteresting because it had to be learnt in a silly way). Today, at any rate, the last step is useless because formal software are wonderful at solving equations, formally or digitally.

About the chemistry of the first of the two phenomena that we have to consider here (dissolving sulfur dioxide in water), we have to start from water on the one hand, and sulfur dioxide on the other.

One could imagine to begin with "Let us assume that we add a mass m of sulfur dioxide to 1 L of water", but, chemically speaking, this requires discussion because in ambient conditions, sulfur dioxide produced by burning sulfur is a gas. In order to put it in solution, one possibility is to store the gas and water in the same vessel, so that sulfur dioxide dissolves according to the Henry model (McQuarrie and Simon, 1997), that states that the concentration in the solution is proportional to the partial pressure in the gas over the solution:

$$[SO_2(sol)] = K \cdot P_{SO_2}$$

where the brackets $[]$ indicates a concentration, K is the Henry constant, and P_{SO_2} is the partial pressure in sulfur dioxide in the gas over the solution.

When sulfur dioxide is dissolved, it can react with water according to the equilibrium:

$$SO_2(sol) + H_2O \rightleftharpoons H_2SO_3$$

The sulfurous acid being formed can give rise to other species:

$$H_2SO_3 + H_2O \rightleftharpoons H_3O^+ + HSO_3^-$$

$$HSO_3^- + H_2O \rightleftharpoons H_3O^+ + SO_3^{2-}$$

Of course, the dissociation of water should always be considered:

$$2H_2O \rightleftharpoons H_3O^+ + OH^-$$

In quantitative terms, we can calculate the pH if we express "conservation" equations:

- the conservation of the mass;

- the conservation of chemical species (atoms...);

- the conservation of energy: one has to remember that this means using the $pKas$, because they are related to chemical potential, the latter being defined as the ratio of the energy by the number of moles;

- the neutrality of the solution (conservation of the electric charge).

In the particular conditions of the question studied here, we have the following equations:

0. Initial condition: the Henry equation for the determination of dissolved sulfur dioxide.

1. Conservation of energy for the dissociation of water (pKe), corresponding to:

$$eq1 := Ke = H \cdot OH :$$

Let us observe that we do not use the brackets denoting concentrations in this equation because such symbols have a special meaning in *Maple*; there are possibilities for using them as we do usually in chemistry, but here we decided to use instead italics to make the difference between objects (compounds, in roman) and their concentration (italics), and we also do not indicate the charges in the mathematical part of the resolution.

2. First equilibrium (conservation of energy, *pK1*):

$$eq2 := K1 = \frac{HSO3 \cdot H}{H2SO3} :$$

Here, H stands for the concentration in protons H$^+$, and HSO_3 stands for the concentration in HSO$_3^-$. You see that we do not use indexes or superscripts (we could, but it would make the work more complex).

3. Second equilibrium (conservation of energy, *pK2*):

$$eq3 := K2 = \frac{SO3 \cdot H}{HSO3} :$$

4. Conservation of electrical charge:

$$eq4 := H = 2 \cdot SO3 + HSO3 + OH :$$

5. conservation of mass:

$$eq5 := H2SO3 + HSO3 + SO3 = c :$$

Now *Maple* can be used to solve the system of five equations and determine the pH.

For the second phenomena, *i.e.*, the oxidation of sulfur dioxide into sulfuric acid, we can assume the general reaction:

$$SO_2 + 1/2O_2 + 2H_2O \rightarrow SO_4^{2-} + 2H^+.$$

As we know the initial amount of SO_2, we can calculate the pH, considering the case of the strong acid that is sulfuric acid. Here, the concentration in protons [H$^+$] is twice the concentration in sulfate ions [SO$_4{}^{2-}$], and the pH can be directly calculated as:

$$pH = -\log10([H^+]) = -\log10(2[SO_4^{2-}])$$

6.5.5 Expressing the Results

6.5.5.1 Finding Digital Data

The dissociation constant of water is known usually given (Lide, 2005):

$$pKe = 14.$$

For the first of these two acid/base equations concerning the dissociation of H_2SO_3, $pK1 = 1.81$ (Sigma-Aldrich, 2018); for the second, $pK2 = 6.99$.

The mass of sulfur dioxide added to wine can be found in data bases (OIV, 2018). Let us admit 100 mg/L, which means a number of moles per liter (mol/L):

$$c = \frac{\left(\frac{0.1}{64}\right)}{1} = c = 1.56 \times 10^{-3}$$

6.5.5.2 Introduction of Data in the Formal Solution

Here, we solve digitally the equations. For the first step:
  $restart$:

$$eq1 := 10^{-14} = H \cdot OH:$$

$$eq2 := 10^{-1.81} = \frac{HSO3 \cdot H}{H2SO3}:$$

$$eq3 := 10^{-6.99} = \frac{SO3 \cdot H}{HSO3}:$$

$$eq4 := H2SO3 + HSO3 + SO3 = 1.56e - 3:$$

$$eq5 := H = 2 \cdot SO3 + HSO3 + OH:$$

Finding the solution can be obtained simply with the *solve* command:

$$solve(\{eq1, eq2, eq3, eq4, eq5\})$$

$\{H = 0.001428388726, H_2SO_3 = 0.0001317135953,$
  $HSO_3 = 0.001428184090, OH = 7.000895355\ 10^{-12},$
  $SO_3 = 1.023146391\ 10^{-7}\}, \{H = -3.205078007\ 10^{-12}, H_2SO_3 = 1.011149588\ 10^{-17},$
  $HSO_3 = -4.886262623\ 10^{-8}, OH = -0.003120048866, SO_3 = 0.001560048863\},$
  $\{H = -2.046483809\ 10^{-7}, H_2SO_3 = -4.122836082\ 10^{-8}, HSO_3 = 0.003120238241,$
  $OH = -4.886430059\ 10^{-8}, SO_3 = -0.001560197012\},$
  $\{H = -0.01691635026, H_2SO_3 = 0.01847645259, HSO_3 = -0.01691655493,$
  $OH = -5.911440614\ 10^{-13}, SO_3 = 1.023305372\ 10^{-7}\}$

In this list of possible solutions, we keep only the one with positive values (the first one). We take the value for "H", *i.e.*, the concentration in protons [H+], and we can calculate the pH:

$$pH = -\log10(0.001428388726)$$

$$pH = 2.85 \times 10^0$$

For the second step, we now have to calculate the pH of a solution of sulfuric acid with the above concentration.

$$pH = -\log10(2[SO_4^{2-}])$$

So that:

$$pH = -\log 10 \left(\frac{2 \cdot 0.1}{64} \right)$$

$$pH = 2.51 \times 10^0$$

6.5.5.3 Discussion

We now have to remember the assumption about the pH of wine before sulfur dioxide is added: we did not consider wine but rather water. For wine, the question is much more difficult because there are a lot of acids (tartaric, malic, succinic, acetic), etc. and salts, making an initial pH often below 4. Indeed, our calculations did not consider buffering systems, which are likely to be present in wine.

The discussion about the buffering of wine calls of course for further calculations, introducing more and more complex solutions in order to model wine. However, formal software solve that easily, which is an advantage for exploring complex solutions.

Regardless of whatever buffering can occur, the comparison of the results for steps 1 and 2 shows that some acidity is produced.

6.5.6 Conclusions and Perspectives

In this exercise, we wanted primarily to show how the new software are changing the way one can calculate in chemistry. This is important because being rid of the complexities of calculation, students can focus on the chemical analysis of the problem.

Indeed, we are now, for pH calculation, in the same situation as 50 years ago, when small and cheap pocket calculators were introduced: slide rulers, logarithm tables, and algorithms for finding square roots were all wiped out.

6.6 IMAGINE THAT YOU LIVE IN A DRY COUNTRY, WHERE WATER IS EXPENSIVE. YOU PAY FOR ONE GLASS OF WATER. WHAT IS THE MAXIMUM QUANTITY OF WATER THAT YOU CAN GET IF YOU PAY FOR "ONE GLASS OF WATER"?

In this section, there will be a difference in the structure of the paragraph: the experiment is given at the end, after calculation, because otherwise it would spoil the pleasure of finding the solution to the question.

6.6.1 The Question

This question clearly calls for creativity, and some students find it difficult because they do not see how to escape the obviousness of the question.

But let us come up with ideas. For example, knowing the chemical composition of food, one could recognize that grass contains up to 99% water so that we could buy grass instead of water, with a lot of grass outside the cavity of the glass.

Still, in the context of physical chemistry, one could imagine playing with temperature and contraction of water: we could fill the glass, cool the water in the glass at 4°C (Lide, 2005), and add more water in the space created in the glass.

But this calls for personal inventivity, and it is perhaps safer to be systematic and investigate the initial list of 14 definitions and formulas, looking for possibilities. Here, we would find the definition of surface tension, and this could lead us to remembering the funny experiment to add more water than the level of the glass. It is shown here how to simply calculate this process.

6.6.2   Analysis of the Question

6.6.2.1   *The Data is Introduced*

No data is given, again: you have to "invent" the problem.

6.6.2.2   *Qualitative Model*

Here you are invited to consider that water is added further in a glass that is already full, such as in Figure 6.11.

This possibility is due to forces toward the center of the liquid that prevent it from flowing. Indeed, many courses in physics explain this by the mutual attraction of water molecules, and an imbalance for the molecules of the upper surface (Figure 6.12).

FIGURE 6.11 *Water over the edge of a glass.*

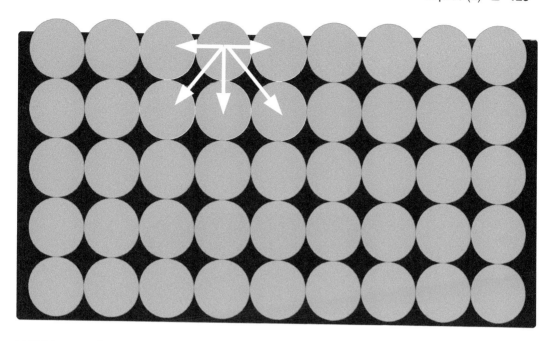

FIGURE 6.12 *A water molecule at the surface is attracted by all other molecules around; however, because there are molecules only inside water, the resultant of forces is toward the center of the liquid.*

For the question studied here, a simple scheme is produced (Figure 6.13). The volume over the glass is considered to be a cylinder.

6.6.2.3 Quantitative Model

The picture being made, one can now introduce the maximum height of water that we can add: h.

Water should also be characterized: here surface tension γ is clearly a good choice for this.

6.6.3 Solving

6.6.3.1 Looking for a Solving Strategy

For this section as for the next one, there will be an exception, because instead of solving the problem, we shall investigate two solutions.

I insist: these "solutions" (indeed, I should only say "proposals") all appear correct, but they are wrong, in different regards, even if the second one gives a result that fits the experimental measurement. We shall discuss very briefly why they are wrong, showing that the Devil is hidden in the details.

The first proposal is to compare the potential energy of elevated water (why the water should flow down) and the surface energy (preventing the liquid from flowing).

The second proposal is to consider pressures in the liquid.

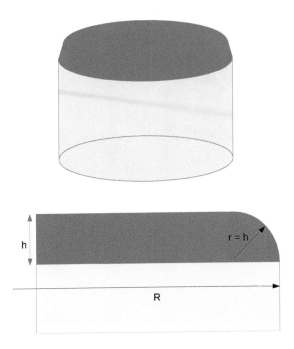

FIGURE 6.13 *A model of the experiment. This calls for a method of making assumptions. In particular, it should be discussed whether the shape of the edge is important, and if the material of the glass should or should not be considered. Of course, in order to simplify the problem, we can assume that r is much smaller than R (radius of the glass).*

6.6.3.2 Implementing the Strategy

6.6.3.2.1 First Wrong Proposal

Water is prevented to flow over the edge by the surface pressure, which is characterized by the "surface tension" γ:

$$\gamma = \frac{\partial G}{\partial a}$$

where G is the Gibbs energy, and a the area. Many students find this formula more difficult than its approximation:

$$\gamma \approx \frac{\Delta G}{\Delta a}$$

For simple calculations, this equation can be used (as long as one remembers that an approximation is made).

Let now be h the maximum height of water that can be added over the top of the edge. Because the upper surface is the same when the glass is simply full or when the level is over the edge, the area increase, Δa, corresponds only to the circular ribbon around water over the edge (assuming that the upper disk moved vertically):

$$\Delta a = (2\pi R) \cdot h.$$

So that:

$$\Delta G = \gamma 2\pi R h$$

Could we say that this energy is equal to the potential energy of water over the edge of the glass? If so, the potential energy for a mass m at height z is

$$m\,g\,z.$$

If we assume that all the mass is in the center of gravity of the slab of water over the edge, the gravity center moved up by $h/2$, hence the potential energy:

$$E = \frac{mgh}{2}$$

The mass of the liquid over the edge is:

$$m = \varrho\pi R^2 h$$

And the energy:

$$E = \frac{\varrho\,\pi\,R^2 h\,g\,h}{2} \xrightarrow{\text{normalize}} E = \frac{\varrho\pi R^2 h^2 g}{2}$$

Thus:

$$\gamma 2\pi Rh = \frac{\varrho\pi R^2 h^2 g}{2}$$

We are looking for h:

$$h = \frac{4\gamma}{\varrho g R}$$

6.6.3.2.2 Second Wrong Proposal

For this second proposal, we keep the same general description (Figure 6.13 top), but we assume a quarter-circle profile at the edge of the glass (Figure 6.13 bottom).

On the figure, the curvature we assumed a radius of curvature of $r = h$ and the angles are certainly not as they should be: for a system with a solid, a liquid and a gas, there is a specific angle that is calculated in physics courses, and this angle is generally not as shown here.

With our wrong assumption, we could use a famous equation for calculating the "Laplace pressure" in curved objects (bubbles of gas, droplets of liquid, meniscus of liquids near glasses, etc.).

This pressure is given by:

$$P = \gamma \cdot \left(\frac{1}{R} + \frac{1}{r}\right)$$

where γ is the surface tension, and R and r the two radii of curvature.

This pressure would make an equilibrium against the hydrostatic pressure, *i.e.*:

$$P = \varrho g r$$

Thus:

$$\gamma \cdot \left(\frac{1}{R} + \frac{1}{r}\right) = \varrho g r$$

And we isolate for r:

$$r = \frac{\gamma + \sqrt{4R^2 g \varrho \gamma + \gamma^2}}{2Rg\varrho}$$

If the radius R of the glass is much large than the radius of curvature of the liquid at the edge:

$$r = \frac{\gamma}{\sqrt{\varrho g \gamma}}$$

6.6.4 Expressing the Results

6.6.4.1 The Formal Result Found is Written Down Again

Here, we simply copy and paste the former results.

With the first wrong proposal, we found:

$$h = \frac{4\gamma}{\varrho g R}$$

You observe that the elevation of the surface of water depends on the radius of the glass, and this is strange because when you carry out the experiment, you do not see such an effect.

And for the second wrong proposal:

$$r = \frac{\gamma}{\sqrt{\varrho g \gamma}}$$

6.6.4.2 Finding Digital Data

We now need the density of water, the size of the glass and the surface tension of water (found in tables). All data should be written in SI units.

Let us use:

- for the density: $\varrho = 1e3$ kg $\cdot$ m^3;

- for the surface tension: $\gamma = 72e - 3$ N $\cdot$ m^{-1} (Hui and Sherkat, 2005).

6.6.4.3 Introduction of Data in the Formal Solution

For the first proposal:

$$subs\left(\varrho = 1e3, r = 0.1, g = 9.8, \gamma = 72e - 3, h = \frac{4\gamma}{\varrho r g}\right)$$
$$h = 2.94 \times 10^{-4}$$

And for the second one:

$$subs\left(\varrho = 1e3, r = 0.1, g = 9.8, \gamma = 72e - 3, \frac{\gamma}{\sqrt{\varrho g \gamma}}\right)$$
$$2.71 \times 10^{-3}$$

In the first case, the increase is about 0.3 mm, which is less than what one can obtain in practice, as we shall see.

And for the second one, the order of magnitude is fine, but the proposal is also wrong.

The analysis of the mistakes that were done would probably take too long to detail in this book, but remember the many assumptions that we made. And there are also many approximations that we did not discuss: for example, we assumed a quarter-circle profile, and we said that this was wrong; however, we also assumed an hydrostatic pressure ϱgh, but this was correct only for the lower part of the liquid over the edge.

The conclusion is not that physics is "difficult", but that we have to learn more and more, to analyze slowly and in more detail; otherwise even the order of magnitude of the results is not obtained. Wonderful physics!

6.6.5 Conclusions and Perspectives

For such an exercise, good advice would be to control the result experimentally, but also to discuss the various parameters in the solution that was found. Of course, it is difficult to change the density of the liquid without changing the surface tension, but adding surfactant can change the surface tension, as shown in the experiment below.

Also, when a calculation is performed, discussing the assumptions is always a good practice.

6.6.5.1 An Experiment to Understand the Question

6.6.5.1.1 First, the General Idea

Now, we know for certain which experiment to do, but it can also be organized as a contest for a group:

1. take a glass;

2. pour some water in it;

3. put it on a scale;

4. with a drop counter, add water drop by drop, and measure the mass at the last drop added before the water flows.

The winner is of course the one who puts more water in the glass.

6.6.5.1.2 Questions of Methods

Experimenting with surface tension is tricky, as shown by the following complementary experiment:

1. take a glass;

2. fill it with very pure water;

3. using the drop counter, add more water until you get a picture, as in Figure 6.11;

4. touch the surface with a very clean metal tip (of a knife or a pin), and observe that nothing occurs;

5. dip the tip in soap, and touch the surface of water again: immediately, water overflows because soap is a surfactant used for reducing the surface tension.

6.6.5.1.3 Where the Beauty Lies

This question serves as a good introduction to surface tension with the many experiments that can be done, making students aware of the presence of the phenomenon in their environment.

One of the most "frightening" experiments is to take brand new capillary tubes (for example 5 μL) from their new box, weigh them and put their lower part in pure water: You will be able to see that the level of water (by capillarity) is very different, in the various tubes. Of course, this could be due to different diameters, for example, so you are invited to weigh the tubes after they are filled with water.

6.7 DOES SYRUP ACCUMULATES AT THE BOTTOM OF A GLASS OF WATER?

In the previous section, we could see how tricky physics can be and why validation, in particular, is so important. Here, however, it will be worse because the experiment will be "bad", and the calculation as well. Beware!

6.7.1 An Experiment to Understand the Question

6.7.1.1 First, the General Idea

For this chapter, we propose to begin with a bad experiment that we shall later improve, after an analysis of its pitfalls.

Indeed, sugar syrups are made mainly of water and sucrose, and we have the "feeling" that sucrose accumulates at the bottom, in glasses of water. But did anyone wait long enough to be sure that an equilibrium is reached?

Let us begin with the simplest experimentation:

1. Fill a glass with water (about 200 g);

2. add a teaspoon of sugar;

3. observe the visible changes as a function of time, up to two days.

We see the sugar crystals falling to the bottom of the glass (this takes seconds), but later, it is not easy to recognize that the sugar crystals have dissolved, making a concentrated solution at the bottom of the glass.

But, as mentioned, this is a poor experiment.

6.7.1.2 Questions of Methods

With sugar alone, phenomena are difficult to observe because sucrose solutions are colorless: one can only see some slight differences in optical density when the glass is moved, as over a hot road in the summer.

Adding a colorant is certainly helpful, and the same experiment as before can be studied using a mint syrup. Of course, the mint syrups that you can make yourself, with the infusion of mint leaves in water, sugar being added later, are colorless, but mint syrups from the industry are green because the industry recognized that they sell better when a green colorant is added (usually, the colorant used is "chlorophyllin", with the same compounds as in plant tissues) (EFSA, 2015).

Another coloring possibility is caramel, and this possibility is also used by the industry: caramel added to food is described as an "additive". There are many different caramels, depending on the production conditions (with sulfides, or ammonium ions, for example), but let us keep it simple:

- heat table sugar in a pan, with a small quantity of water; soon, the content of the pan turns brown, with a caramel odor.

- then add more water and heat a little, to ensure the dissolution of the caramel in water.

- Finally, when a caramel solution is produced, store it and wait, observing at regular intervals.

6.7.1.3 Where the Beauty Lies

Placing sugar crystals at the bottom of a test tube full of water creates a density "gradient" (Fisher and Cline, 1963): the sucrose molecules dissolve in water, and because of molecular diffusion (McQuarrie and Simon, 1997), they distribute slowly from the bottom of the tube toward the other parts of the tube, a process that can be described as a net diffusion of sucrose and water, equilibrating the concentration.

Such sucrose gradients are used for the separation of large molecules (macromolecules, such as proteins, for example) or particles (such as organelles in biology): the material is put at the top of tubes in which a gradient of sugar concentration was prepared, and the tubes are centrifuged. One factor that determines the sedimentation velocity is the difference between the density of the particle and the density of the solvent: if the density of the particle is greater than the density of the solution, at a certain distance from the top, the particle will sediment, but when there is no density difference, there will be no sedimentation, whatever the acceleration.

Continuous gradients are, however, slow to produce, and biochemists prefer discontinuous gradients that are obtained just as when preparing a cocktail known as "Irish Coffee" (with a bottom layer of coffee, a middle layer of whiskey and a top layer of whipped cream). In labs, there are only layers of sugar syrups of decreasing densities from the bottom toward the top.

6.7.2 The Question

Does syrup accumulate at the bottom of a glass of water?

6.7.3 Analysis of the Question

6.7.3.1 The Data is Introduced

Here again, no data is given, and the students have to invent the solution by themselves.

FIGURE 6.14 *First, 5 g of homemade caramel was added to water, before 2 g of a green mint syrup was added. After two days, this appearance is observed, with an accumulation (gradient) of sugar at the bottom of the glass. The caramel and green colorant distribute with different gradients than sucrose, as shown as the appearance of the two visible layers.*

6.7.3.2 Qualitative Model

Figure 6.14 shows what is obtained by adding both mint syrup and caramel to water.

Certainly, adding syrup to water leads to a bottom layer of concentrated syrup, and one needs to stir the mixture to get a homogeneous drink. After some time, the gradient is again established. Doesn't this deserve interpretation and calculation?

When the syrup is added to water, the question of its sedimentation is simple (the density of the syrup is higher than the density of water), and calculating the density of the syrup could be an answer to the question of this chapter; as well, one could calculate the velocity of sedimentation, but this will be studied later (Chapter 9, Section 9.2).

Here, we shall instead focus on the distribution of sucrose when an equilibrium is reached: why is there a bottom layer with more sugar at the bottom of the glass? How would sucrose molecules distribute in the liquid after a lengthy period of rest?

As said, this question is important for separation in biology laboratories, but also for chemical analysis, because imagine that you have to determine the quantity of sucrose in an aqueous solution: without care, you could find different results, depending on the position of the sample that you take. This question is also a way to illustrate the possibilities of mistakes during analyses: sampling is a difficult art that calls for much insight.

For this second question of sucrose distribution in aqueous solution at equilibrium, making a model is a key issue because one has to decide about parameters. The analysis runs as follows: if there is a distribution in the height of sucrose, this means first that gravity is important. And as an equilibrium is to be considered, the question calls for thermodynamics. Students in food science had courses on classical thermodynamics, but also some hints on statistical thermodynamics. And they could probably study the "barometric question", *i.e.*, the variation of pressure in the atmosphere as a function of altitude.

Here, certainly, sucrose molecules in dilute solutions are objects that interact with water molecules, and one could try to use the Maxwell-Boltzmann distribution, but we have to be very careful, because... Why should this physical model apply? The question is important because this distribution was obtained using some particular assumptions that have to be fulfilled so that the equations hold.

6.7.3.3 Quantitative Model

We need to introduce:

- the height above the bottom of the glass h.

- the full height of the glass H.

- the absolute temperature T.

- the molar mass of sucrose molecules M.

6.7.4 Solving

6.7.4.1 Looking for a Solving Strategy

Here the strategy will be very different, because the goal itself is different. We are going to make different calculations and examine why it is good to be cautious about assumptions.

6.7.4.2 Implementing the Strategy

Remember that we had the feeling that the Maxwell-Boltzmann distribution could be the equation (among the 14) that could apply. It includes a ratio between the energy of the considered objects and the kinetic energy from the environment. Here

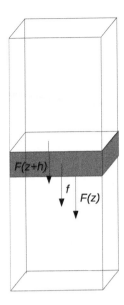

FIGURE 6.15 *The pressure, in a column of air of unit area, is changing with the altitude.*

the kinetic energy, of the order of magnitude of $k_B T$, would be compared to the gravitational potential: we know that this gravitational potential is mgh for a body of mass m at height h in the gravitational field of the Earth, characterized by g.

But for molecules? Applying this model is doubtful because one molecule of sucrose can go in any direction, depending on the interactions with the neighboring molecules (of water). Indeed, if you look at your courses on statistical thermodynamics, you will see that the Maxwell-Boltzmann distribution was established assuming no interaction between the particles, such as in a dilute gas. It does not apply to liquids, for which the interactions are many, in particular with syrups, with sucrose molecules establishing many hydrogen bonds with water molecules.

Could we apply another result of our courses, *i.e.*, the "barometric distribution? In courses on the thermodynamic of atmosphere, and in particular for the calculation of the distribution of pression in function of the altitude (isotherm atmosphere), one would first consider the pressures at altitude z and $z+h$; if ρ (z) is the density of air at this altitude z, we would have:

$$P(z) = P(z + h) + \rho(z)gh.$$

You find this difficult? You are right, because you are probably happier with forces, and you use, as a first idea, the proposal that the pressure is the force applied by unit area.

Let us first make a picture (Figure 6.15), assuming a column of air, of area equal to 1.

The pressure at a certain level ($z+h$, for example), is due to the weight $F(z+h)$ of the air above this level.

Now we understand that, in order to link $F(z)$ and $F(z+h)$, we need to know the weight f of the slab of air in red. It is equal to the mass m of this slab multiplied by g:

$$f = mg$$

Because:

$$\varrho(z) = \frac{m}{v}$$

we see that:

$$f = \varrho(z)vg$$

The volume being equal to $h \times 1 \times 1$, we have:

$$f = \varrho(z)hg$$

Now, and now only, we are ready to move on, from the formula:

$$P(z) = P(z + h) + \varrho(z)gh.$$

This can be rearranged as:

$$\frac{P(z + h) - P(z)}{h} = -\varrho(z)g$$

And the limit, when h tends to 0, is:

$$\frac{dP(z)}{dz} = -\varrho(z)g$$

Imagine that air is an ideal gas of constant temperature (which is quite unrealistic). Then, the ideal gas equation would hold:

$$PV = nRT$$

But this equation can be made in relation with the density:

$$\varrho(z) = \frac{m}{V} = \frac{P(z)m}{n}\frac{1}{RT} = \frac{P(z)M}{RT}$$

where $M = \frac{m}{n}$ is the molar mass of air.

Now, the differential equation can be written:

$$\frac{dP(z)}{dz} = -P(z)\frac{Mg}{RT}$$

with $P(0) = P_0$.

Here, the students of the past had to apply resolution methods, such as variable separation (you group all the P in one side), but using *Maple* or an equivalent software makes the resolution faster:

$$dsolve\left(\left\{\frac{d}{dz}(P(z)) = -P(z)\frac{Mg}{RT}, P(0) = P0\right\}, P(z)\right)$$

$$P(z) = P0\,e^{-\frac{Mgz}{RT}}$$

This being done, forget about this barometric question, and come back to our initial question, taking the worst approach: applying the Maxwell-Boltzmann when we should not. Let us assume that each molecule of sucrose is as a small object of mass m and volume v, with forces acting on it (weight, buoyancy), in a continuous medium that is water.

The sum of forces related to gravity (weight, buoyancy) on sucrose molecules would be:

$$R = mg - \varrho_w v g$$

where g is the acceleration of gravity, ϱ_w the density of water.

Let us observe immediately that using this sum, we do not take into account any "friction", such as the Stokes force $6\pi\eta V r$, with η the viscosity of the fluid, V the velocity of the particles and r their radius (Stokes, 1851).

Nevertheless, if we express the volume v as a function of the density ϱ_p of the molecule through $m = \varrho_p v$, the sum of forces would be:

$$R = mg - \varrho_w \left(\frac{m}{\varrho_p}\right) g = mg \left(1 - \frac{\varrho_w}{\varrho_p}\right)$$

Now, we could calculate the potential in which the sucrose molecule is staying: it is the opposite of the work of forces of gravity, which is expressed as:

$$\vec{R} = -\overrightarrow{grad V}$$

So that the potential V is:

$$V = m \cdot g \cdot h \left(1 - \frac{\varrho_w}{\varrho_p}\right)$$

According to the Maxwell-Boltzmann distribution applied boldly, the proportion of molecules at height h would equal to:

$$d(h) = A \exp\left(-\frac{mgh\left(1 - \frac{\varrho_w}{\varrho_p}\right)}{kT}\right)$$

where A is a constant that can be found using the normalization expression:

$$N = \int_0^H d(h) dh = \int_0^H A \exp\left(-\frac{mgh(1 - \frac{\varrho_w}{\varrho_p})}{kT}\right) dh = A \int_0^H \exp\left(-\frac{mgh(1 - \frac{\varrho_w}{\varrho_p})}{kT}\right) dh$$

As:

$$\int_0^H \exp\left(-\frac{mgh\left(1 - \frac{\varrho_w}{\varrho_p}\right)}{kT}\right) dh$$

is equal to:

$$\frac{\varrho_p kT \left(e^{-\frac{mgH(\varrho_p - \varrho_w)}{\varrho_p kT}} - 1\right)}{-mg(\varrho_p - \varrho_w)}$$

We would find:

$$A = \cfrac{N}{-\cfrac{\varrho_p kT\left(e^{-\frac{mgH(\varrho_p-\varrho_w)}{\varrho_p kT}}-1\right)}{mg(\varrho_p-\varrho_w)}}$$

And finally:

$$d(h) = \cfrac{N}{-\cfrac{\varrho_p kT\left(e^{-\frac{mgH(\varrho_p-\varrho_w)}{\varrho_p kT}}-1\right)}{mg(\varrho_p-\varrho_w)}}\ \exp\left(-\cfrac{mgh\left(1-\frac{\varrho_w}{\varrho_p}\right)}{kT}\right)$$

The shape of this variation is a decreasing exponential, as for the barometric equation but remember that all this was done when it should not have been. The fact that the shape of the distribution is reasonable is not a demonstration that the calculus holds. And this is why it is so important to test theories experimentally: as Richard Feyman said, "In general, we look for a new law by the following process: First we guess it; then we compute the consequences of the guess to see what would be implied if this law that we guessed is right; then we compare the result of the computation to nature, with experiment or experience, compare it directly with observation, to see if it works. If it disagrees with experiment, it is wrong. In that simple statement is the key to science. It does not make any difference how beautiful your guess is, it does not make any difference how smart you are, who made the guess, or what his name is — if it disagrees with experiment, it is wrong."

6.7.5 Expressing the Results

6.7.5.1 The Formal Result Found is Written Down Again

Let us write again the former final expression:

$$d(h) = \cfrac{N}{-\cfrac{\varrho_p kT\left(e^{-\frac{mgH(\varrho_p-\varrho_w)}{\varrho_p kT}}-1\right)}{mg(\varrho_p-\varrho_w)}}\ \exp\left(-\cfrac{mgh(1-\frac{\varrho_w}{\varrho_p})}{kT}\right)$$

We see a constant value before an exponential of the form $\exp(-Kh)$. The recognition of this shape will be used later, in the discussion.

6.7.5.2 Finding Digital Data

In the expression above, we need first to fix the size (height) of the glass: 0.1 m.

For N, let us consider 10 g of sugar (the molar mass for a molecule of formula $C_{12}H_{24}O_{12}$ is $M = 12 \cdot 12 + 24 + 12 \cdot 16 = 360$), so that the number of moles is 10/360, and the number of molecules

$$\frac{10}{360} \cdot 6 \cdot 10^{23} = 1.67 \times 10^{22}.$$

For the mass (in kg) of one molecule, this is the molar mass divided by the number of moles, i.e.:

$$m = \frac{360e-3}{6 \cdot 10^{23}} = 6.10^{-25}.$$

The density of water is known: 1000 kg/m^3.

But the question of the "density" of the sucrose molecules (remember, not of syrup) is a difficult one because it calls for taking into account the interactions of sucrose and water molecules; in particular because of many hydrogen bonds. One solution could be to look on the Internet for molecular models, but such molecular modeling would be based on models of water and there are many: which one to choose?

In order to make an educated guess, let us imagine that we use the previous result (Chapter 6, section 6.1), about the distance between water molecules in liquid water: we found an average distance of about 3×10^{-10} m, *i.e.*, the double of a covalent bond.

Using a software for chemistry (Figure 6.16), we can find a (van der Waals) volume of 0.3 nm^3. This would give a density of:

$$\frac{6e-25}{0.3e-27} = 2.0 \times 10^3, \text{ double that of water density.}$$

6.7.5.3   Introduction of Data in the Formal Solution

For H, let us assume the full height of a glass.

We calculate the distribution for two heights: For $h = 0.1$:

$$evalf\left(subs\left(H = 0.1, g = 9.8, k = 1.38e-23, T = 293, m = 6e-25, h = 0.1,\right.\right.$$

$$\varrho_w = 1e3, \varrho_p = 2e3, N = 1.7e22, -\frac{N}{\dfrac{\varrho_p kT\left(e^{-\frac{mgH(\varrho_p-\varrho_w)}{\varrho_p kT}}-1\right)}{mg(\varrho_p-\varrho_w)}}\exp\left(-\frac{mgh\left(1-\frac{\varrho_w}{\varrho_p}\right)}{kT}\right)\right)$$

$$1.70 \times 10^{23}$$

For $h = 0.01$:

$$evalf\left(subs\left(H = 0.1, g = 9.8, k = 1.38e-23, T = 293, m = 6e-25, h = 0.01,\right.\right.$$

$$\varrho_w = 1e3, \varrho_p = 2e3, N = 1.7e22, -\frac{N}{\dfrac{\varrho_p kT\left(e^{-\frac{mgH(\varrho_p-\varrho_w)}{\varrho_p kT}}-1\right)}{mg(\varrho_p-\varrho_w)}}\exp\left(-\frac{mgh\left(1-\frac{\varrho_w}{\varrho_p}\right)}{kT}\right)\right)$$

$$1.70 \times 10^{23}$$

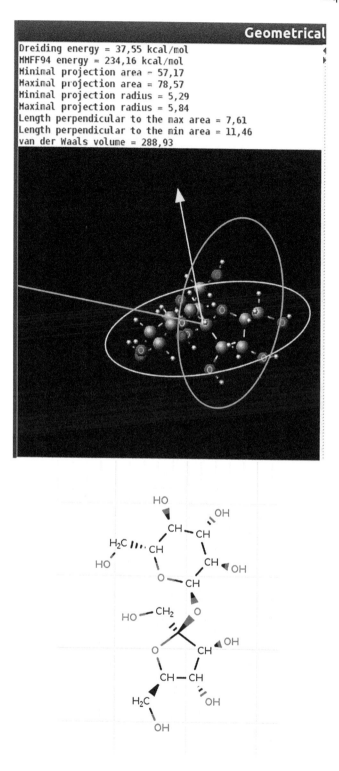

FIGURE 6.16 *The molecular representation of sucrose (top), and the determination of the geometrical parameters using the software Marvin (bottom): the results are in angströms.*

We observe that we get the same result. How is it possible?

6.7.5.4 *Discussion*

All that has been done previously should be discussed carefully. Certainly, the last calculation about the "density" of a molecule is too hasty. But the main point for discussion is about the possibility or not of using the Maxwell-Boltzmann distribution.

Indeed, the possibility is doubtful because these statistics were built assuming that there are no interactions between particles; and, as said, it only holds, strictly speaking, for ideal gases. It can be used as an approximation in the case of real gases when the interaction between particles can be neglected but it cannot be used for liquids, for example.

For colloids, the sedimentation equilibrium was calculated by Jean Perrin (1870–1942), Nobel prize winner in physics in 1926 (Perrin, 1913). The equilibrium is reached when the sedimentation flux of molecules is exactly compensated by the flux of matter because of kinetic energy. Here the Mason-Weaver equation describes the sedimentation and diffusion of solutes (here sucrose) under the gravitational field:

$$\frac{\partial}{\partial z}c(z,t) = D\frac{\partial^2}{\partial z^2}c(z,t) + sg\frac{\partial}{\partial z}c(z,t)$$

where z is the direction, t the time, c the concentration; the parameters D, s, g represents the solute diffusion constant, the sedimentation coefficient, and the acceleration of gravity. At equilibrium, the concentration distribution agrees with the Maxwell-Boltzmann distribution.

However, for our question, sucrose is certainly not a colloidal particle, and one could ask whether the theory applies. Let us remember that the profile that we calculated was of the form $A \cdot \exp(-\frac{h}{L})$, with a characteristic length L equal to $\frac{k_B \cdot T}{m \cdot g \cdot (1 - \frac{\varrho_w}{\varrho_p})}$

Numerically, this length is equal to about 10^3, *i.e.,* about 1 km: clearly, this should not have any visible effect in the experimental conditions studied.

6.7.6 Conclusions and Perspectives

This exercise shows that one should be careful in applying physical formulas: one has to learn them along with the conditions of their application. And it places the students closer to modern questions of research, showing in particular the interest of molecular modeling.

Finally, you are invited to continue the discussion by reading the article of Kaufman and Dorman (2008): even the simple idea that chemists have of dilute solutions should be carefully discussed.

6.8 BY HOW MUCH DOES COFFEE COOL BY RADIATION?

6.8.1 An Experiment to Understand the Question

6.8.1.1 First, the General Idea

Hot food can cool by radiation, conduction, convection, and evaporation. Indeed convection was said to be discovered by the American Benjamin Thomson (1753–1814), later knighted as count Rumford, because of his soup: once, he had a thick soup, and he burned his mouth because there was no convection, contrary to other more liquid soups that he had before; for these, only the upper surface cooled, and the other parts below remained hot (Kurti, 1969). Rumford also studied the transformation of mechanical work into heat, when piercing canons.

With coffee, conduction occurs, and this can be easily be perceived: the cup heats. Convection also occurs because the upper surface cools faster than the bottom, so that the cooler upper coffee becomes denser than the bottom, hotter liquid, that flows down when hotter lower coffee moves up. Radiation? Remember the experiment of putting your hand very close to one of your cheeks without touching it, and you feel "heat": this occurs also for the coffee and the cup. And evaporation of water can be seen in the "smoke" over the coffee: this means that some water evaporates and recondenses when arriving in cooler air, above the cup: indeed, we do not see the vapor, but rather the water droplets that form from it.

But the question is: how much of each process?

6.8.1.2 Questions of Methods

Traditionally, temperature measurements were performed using thermometers (with alcohol, or mercury), but one can find much better devices with thermocouples; and one success of molecular cooking is the introduction of such devices in kitchens all over the world. They are precise, they don't break, you can put them into ovens, and you can make them yourself (Kurti and This-Benckhard, 1994): you simply needs to wield two metallic wires of different natures, such as iron or copper, or more often chromel (alloy of nickel and chromium) and alumel (nickel and aluminum, plus some silicon). At the free ends, an electric difference of potential that can be measured depends on the temperature of the junction.

For the measurements, one needs a multimeter that measures the voltage resulting from the so-called Seebeck effect (discovered in 1821 by the German physicist Thomas Johann Seebeck), a result of the fact that at the junction, electrons are more attracted by the nucleus of one metal than by the other.

Of course, if you make your own thermocouple instead of buying one, you will need to calibrate it, measuring a difference of electrical potential first in a mixture of ice and water (V_1), and then in boiling water (V_2); then you can interpolate between these two points of coordinate $(0, V_1)$ and $(100, V_2)$. But you can also use a traditional thermometer to create a table with corresponding voltages as a function of temperature.

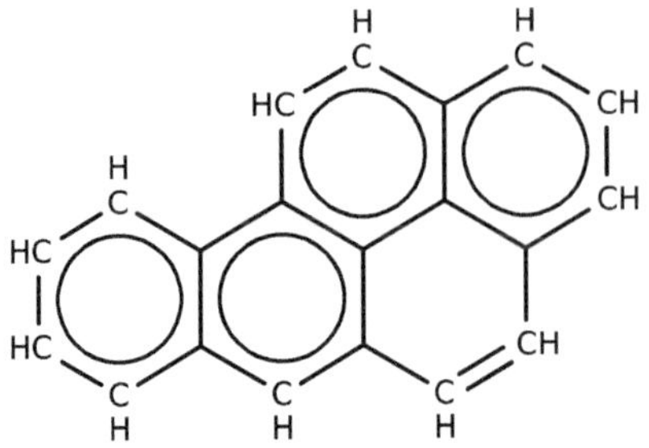

FIGURE 6.17 *Benzo[a]pyrene is one member of the benzopyrene family, polycyclic aromatic hydrocarbons that are produced during incomplete combustion of aromatic compounds, such as lignin from wood. They have different toxicities but are often cancerogenic, because they can stack between the base pairs of DNA and perturb replication (EFSA, 2008b).*

6.8.1.3 Where the Beauty Lies

Radiation, *i.e.*, the propagation of electromagnetic waves, was a great mystery, at the time of Michael Faraday, who had discovered in 1845 that electricity and magnetism could interact with light: the polarization of light turns by an angle proportional to the magnetic field along the direction of propagation (Tyndall, 1868). But it needed the mathematical talent of James Clerk Maxwell for translating the ideas of Faraday into a mathematical description, with four equations describing the propagation of electromagnetic waves (Maxwell, 1855). Faraday, however, did not stop his research with the discovery of the relationship of electricity, magnetism, and light: convinced that there was some "unity" in the world, he also looked for a relationship with gravity. This was to be found only much later, with general relativity.

Almost two centuries after Faraday, there is still a frequent misconception about radiation, and too many people eat meats loaded with cancerogenic benzopyrenes (Figure 6.17), produced with barbecues, because they have the feeling that the meat should be *over* the fire or the embers, where the smoke is depositing the benzopyrene and other toxic compounds: they forget that in their ovens the heating resistance of the grill is over the food. Thus, there has been a decrease in culinary knowledge; in the last century, cooks well knew that infrared moves in all directions without being modified by gravity; as a result, they were placing the meat in front of the fire, with a hemispheric canister–acting as a reflector–placed behind the meat, and a drip pan under the meat, for juice recovery.

6.8.2 The Question

In Section 5.3 of Chapter 5, we proposed an exercise about the Devil radiating like a black body, a result which students cannot easily apply in the kitchen. Here, the question is almost the same but it is more practically applicable.

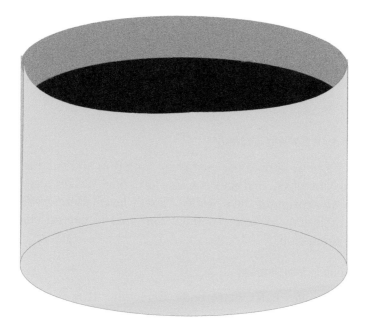

FIGURE 6.18 *Coffee cooling in a cup. After the simple representation, the question has to be analyzed word by word.*

This new question has another interesting point: frequently, in educational circumstances, different mechanisms are proposed in order to explain phenomena, but the issue is to find the most important mechanism. Orders of magnitudes are important for many reasons, but in particular because they help determine which mechanism is more important than others. Here we have an example for which calculations are useful for deciding as well.

6.8.3 Analysis of the Question

6.8.3.1 The Data is Introduced

Again no data is given.

6.8.3.2 Qualitative Model

A simple picture can be made, as shown in Figure 6.18. Coffee cools by various mechanisms: conduction from the liquid to the cup, conduction from the cup to the air, convection in the coffee, convection in the air, evaporation of water, and radiation by the upper surface and from the cup.

6.8.3.3 Quantitative Model

For the simplest model, we consider only the drink:

- there is a mass m of coffee;

- the initial temperature is T_1.

- the final temperature is T_2.

For a more complex system, the cup could be considered as well.

6.8.4 Solving

6.8.4.1 Looking for a Solving Strategy

Generally, solid physical bodies lose energy by radiation and conduction, but liquids can also undergo convection, and they can also lose some of their mass by evaporation, which needs the absorption of the so called "latent heat". Here, we focus only on the energy lost by radiation, in order to show that this is not an efficient process of cooling (see the discussion).

6.8.4.2 Implementing the Strategy

Let us consider the coffee like a black body. Remember that it radiates a power:

$$P = \varepsilon \cdot \sigma \cdot A \cdot T^4$$

where ε is the emittivity (1 for a black body), A is the area of the emitting body, σ the Stefan-Boltzmann constant, T the absolute temperature (varying with time t). Here, let us assume that the emittivity is 1.

The energy dE given by the coffee during a time interval dt is obtained through:

$$P(t) = \frac{dE}{dt}$$

Here, you see that the power is a function of t because when time goes on, the temperature decreases the energy itself is changing with time: when time goes on, the temperature decreases. We can see this more clearly in:

$$dE(t) = \sigma \cdot A \cdot T(t)^4 \, dt$$

Let m be the mass of coffee (we could play with a changing mass as well through evaporation, but let us assume here that it will be constant). The decrease in temperature

$$dT(t)$$

is related to exchanged heat $dQ(t)$ by:

$$dQ = m \cdot c_p(t) \cdot dT(t)$$

where $c_p(t)$ is the heat capacity of coffee (we assume that it behaves as water, from this point of view). So that

$$A \cdot \sigma \cdot T(t)^4 \cdot dt = m \cdot c_p(t) \cdot dT(t)$$

Some students can find it difficult to solve such an equation, but it is a good thing that they find a solution anyway, even an approximated one. One simplification is to overestimate the radiation, by considering that the coffee is radiating always as for temperature T_1.

With this assumption, solving for a constant temperature is easier because the heat capacity is constant as well. We introduce the time τ for radiation:

$$E = P \cdot \tau = A \cdot \sigma \cdot T^4 \tau = m \cdot c_p \cdot \Delta T$$

From which we find the time τ:

$$\tau = \frac{m \cdot c_p \cdot \Delta T}{A \cdot \sigma \cdot T^4}$$

Here we are.

More advanced students could consider that between the times t and $t + dt$, the radiated energy is $A \cdot \sigma \cdot T(t)^4 dt$, taking the variation of T into account. And a first step is to keep the assumption that c_p is constant. This means that the equation to solve becomes:

$$m \cdot c_p \cdot (T(t + dt) - T(t)) = A \cdot \sigma \cdot T^4 dt$$

Some algebraic manipulations can give the variation in temperature as a function of time:

$$(T(t + dt) - T(t)) = \frac{A \cdot \sigma}{m \cdot c_p} \cdot T(t)^4 \cdot dt$$

$$dT(t) = \frac{A \cdot \sigma}{m \cdot c_p} \cdot T(t)^4 \cdot dt$$

This is a differential equation that can be solved by integrating two independent parts, as can be seen if one writes:

$$\frac{dT(t)}{T(t)^4} = \frac{A \cdot \sigma \cdot dt}{m \cdot c_p}$$

$$\int_{T_1}^{T_2} \frac{1}{T^4} dT = \int_0^\tau \frac{A \cdot \sigma}{mc_p} dt$$

$$assume(0 < T_1 < T_2)$$

For the first integral:

$$\int_{T_1}^{T_2} \frac{1}{T^4} dT$$

$$-\frac{1}{3} \frac{T\sim_1^3 - T\sim_2^3}{T\sim_1^3 \, T\sim_2^3}$$

And for the second integral:

$$\int_0^\tau \frac{A \cdot \sigma}{mc_p} dt$$

$$\frac{A\sigma\tau}{mc_p}$$

Coming back to the equation:

$$eq := -\frac{1}{3}\frac{T\widetilde{~}_1^3 - T\widetilde{~}_2^3}{T\widetilde{~}_1^3\, T\widetilde{~}_2^3} = \frac{A \cdot \sigma \cdot \tau}{mc_p}$$

$$eq := -\frac{T_1^3 - T_2^3}{3T_1^3 T_2^3} = \frac{A\sigma\tau}{mc_p}$$

We look for the time τ:

$$solve(eq, \tau)$$

$$-\frac{(T_1^3 - T_2^3)\, mc_p}{3T_1^3 T_2^3 A\sigma}$$

6.8.4.3 Validation

Here, the second solution, being an approximation of the first, is not really a validation but one way to explore the result.

6.8.5 Expressing the Results

6.8.5.1 The Formal Result Found is Written Down Again

For the first case:

$$\tau = \frac{mc_p \Delta T}{A \cdot \sigma T^4}$$

For the second case:

$$\tau = -\frac{1}{3}\frac{(T\widetilde{~}_1^3 - T\widetilde{~}_2^3)mc_p}{T\widetilde{~}_1^3\, T\widetilde{~}_2^3\, A \cdot \sigma}$$

6.8.5.2 Finding Digital Data

Let us assume a mass of coffee of 100 g (0.1 kg).

We decide to calculate the energy loss between the initial temperature (assumed to be 100°C) and the final temperature, when it is not too hot to be consumed, *i.e.*, 40°C.

We have to translate Celsius degrees (°C) in kelvins (K).

The other constants were introduced before in this book.

6.8.5.3 Introduction of Data in the Formal Solution

$$subs\left(m = 0.1, c_p = 4.16e3, \Delta T = 60, A = 1e-3, \sigma = 5.6703e-8, T = 300, \right.$$

$$\left. \tau = \frac{mc_p \Delta T}{A \cdot \sigma T^4} \right)$$

$$t = 5.43 \times 10^4$$

This is in seconds; with more intuitive units, it makes about 33 hrs.
    For the second case:

$$subs\left(m = 0.1, c_p = 4.16\mathrm{e}3, \Delta T = 60, A = 1\mathrm{e} - 3, \sigma = 5.6703\mathrm{e} - 8,\right.$$

$$\left. T_1 = 373, T_2 = 273 + 40, \tau = \frac{1}{3}\frac{(T\tilde{}_1^3 - T\tilde{}_2^3)mc_p}{T\tilde{}_1^3\, T\tilde{}_2^3\, A\,\sigma}\right)$$

$$t = 3.26 \times 10^4$$

This is of the same order of magnitude, but of course it is different because of our differing assumptions.

6.8.5.4 *Discussion, Playing with Parameters in Order to Explore the Solution Space*

The discussion is within the choice of validation: the two different solutions are one way to establish that cooling by radiation is a quite inefficient process. Obviously, conduction, convection, and evaporation (remember: your soup cools faster when you blow over it) are important.

6.8.6 Conclusions and Perspectives

Here, we did not solve the two conduction and convection cases. For the first case, the Fourier model is to be used. For the second, one could use the Newton equation. And finally, a cup of coffee could be weighed before and after cooling in order to estimate how much heat is lost by evaporation; of course, a calculation also can be made using the vapor pressure as a function of the temperature.
    In practice, one should be aware that coffee stirring is much more efficient because it increases the exchange surface between the hot liquid and cold air. In particular, if you blow on your coffee when you stir it, you will displace the fastest water molecules in the vapor phase, making a generally colder liquid. This is so important that it was the basis of the introduction of the famous course on physics by Richard Feynman (Feynman *et al.*, 1963).

6.9 WHEN AN EGG (WHOLE EGG, IN THE INTACT SHELL) IS DIPPED IN VINEGAR, WHY DOES IT BECOME BIGGER AFTER A NUMBER OF DAYS?

6.9.1 An Experiment to Understand the Question

6.9.1.1 *First the General Idea*

The experiment is so easy and cheap that it is good to try:

1. take a bowl;

2. fill it with spirit vinegar (certainly, this is not interesting from a culinary point of view, but being colorless, the phenomena will appear more clearly);

3. observe the phenomena at regular interval over two days;

4. when the entire shell is dissolved and you observe the beginning of a coagulation, add 100 g of salt or sugar and observe the shrinking of the egg, within minutes;

5. pour off the liquid and replace it with water: the egg will swell again.

The swelling and expansion of eggs depending on their liquid environment is a result of osmosis, a phenomenon caused by the different affinity of water inside and outside the egg. Often, courses in physics and physical chemistry speak of "semipermeable membranes", and students hardly imagine them, but here, the membrane of the egg is one example; water is almost the only substance that can go through (but it is not the only one, and the proof is given below).

6.9.1.2 Questions of Methods

Here, it is convenient to go deeper into the experiment, trying to leave no phenomenon without interpretation.

Let us begin with eggs: chicken eggs have an oval shape, with one end rounded and the other more pointed. This shape results from the egg being forced through the oviduct.

Their color? Being mainly made of white calcium carbonate $CaCO_3$, the shell of eggs should be white, but some varieties of chicken have colored eggs; in particular, protoporphyrin compounds (related to the red hemoglobin) produce reds and browns as a ground color or as spotting (Milgrom, 2001).

The mass? About 60 g, even if there are variations, and indeed eggs can give an order of magnitude of masses in the kitchen.

Their composition? The shell, the white, and the yolk; the latter floats in the white, contrary to what most people think, and this occurs because the main difference in composition is the presence of lipids in the yolk. How to be sure of it? There are different ways:

- holding the egg with the long axis vertical, use a sharp knife to cap off, and you will be able to observe the yolk at the top;

- putting many egg whites and one egg yolk in a test tube of about 5 cm diameter or in a narrow glass, you will see that the yolk is floating;

- since some people think that these two experiments do not demonstrate anything because they disrupt the "chalazes" (membrane cords on each side of the yolk), there are other experiments such as the following:

  - put an egg in a pan with water, make a cross with a pencil on the upper part, and boil for 10 min, avoiding rolling the egg; when you cut it, you will be able to observe the position of the yolk at the top;

- and the last demonstration will be given by this experiment here, with vinegar.

Now, let us put this egg in a transparent vessel, and add vinegar: in some seconds, the shell becomes white when it had initially another color (pink or brown). If you look carefully, you will see that the surface of the egg is covered with tiny bubbles that expand and detach, going toward the surface of the liquid: there is a chemical reaction between the acetic acid of the vinegar and calcium carbonate, producing carbon dioxide, while calcium ions go in solution.

After about two hours (it depends on the strength of the vinegar), there is no shell left, and if you used a colored egg, a red-brown membrane can be observed floating; you also see the yolk floating inside the egg white, which is the last demonstration announced above.

If you wait longer after storing eggs in vinegar, you will see that the egg expands generally; hence the question for this chapter. And finally, after a number of weeks, the egg white coagulates, as well as the yolk. This is why it was said above that not only water can move through the membrane; however, it is true that it is much faster for water to do so than other compounds such as acetic acid.

6.9.1.3 *Where the Beauty Lies*

As mentioned, if you wait many months, after storing eggs in a strong vinegar (*i.e.*, a vinegar whose content of acetic acid is high), you will see that the white gets opaque and that the yolks even coagulate strangely. Indeed, such eggs stored in vinegar do not spoil: I have kept some for more than 20 years in my laboratory. They can be used as hard boiled eggs in a salad, but with a different flavor (for this application, preferably use a good vinegar).

A name for such eggs? Because I had the idea of them after analyzing the "one century old eggs" from Asia (produced by storing eggs in basic mixtures containing lime or potash), and I used acids instead of alkalis, I called them "minus one century old eggs".

By the way, do not throw away the vinegar because it contains calcium ions Ca^{2+}, that can find an application for making firm jams: traditionally, jam makers used copper for that, and it is true that copper ions Cu^{2+} contribute to make firmer jams, holding two pectin molecules together, through electrostatic bonds with the carboxylate ions CH_3COO^-, but copper in high quantities is not safe, and divalent calcium is preferable (Schell *et al.*, 2012).

6.9.2 The Question

When an egg is stored in vinegar and its shell is dissolved, then the egg expands, but it can shrink if a lot of salt is added to the vinegar. Why?

6.9.3 Analysis of the Question
6.9.3.1 *Data is Introduced*

The mass of an egg is about 60 g, and the shell is about 10% of it (Belitz *et al.*, 2009).

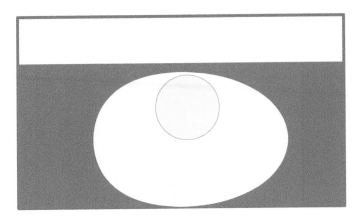

FIGURE 6.19 *Storing an egg in vinegar.*

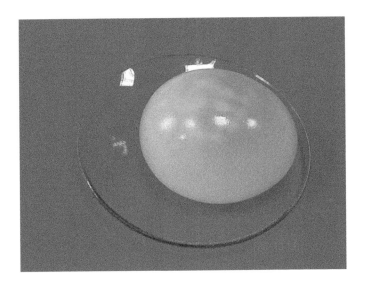

FIGURE 6.20 *An egg after its shell was dissolved in a vinegar.*

The egg white can be considered to be half of the egg, *i.e.*, 30 g (OK, we could say 2/3 as well, but it is the same order of magnitude), and it is made of 90% water, and 10% proteins.

The egg yolk can be considered to be half of the egg, *i.e.*, 30 g, and it is made of 50% water, 35% lipids, and 15% proteins.

For vinegar, it is typically a solution of 5–8% (w/w) of acetic acid.

6.9.3.2 Qualitative Model

Let us begin with a picture of the experiment (Figure 6.19), as usual (because it tells us how to make the qualitative model: we describe what we see.

And now here is the result (Figure 6.20).

The assumptions are:

- the vinegar can be assumed to be an aqueous solution of acetic acid (remember CH_3COOH).

- the shell is mainly made of calcium carbonate ($CaCO_3$).

- the egg white contains proteins (see data).

6.9.3.3 Quantitative Model

For the vinegar, p is the mass proportion of acetic acid.

For the egg, the mass is M, with a mass M_s for the shell, and M_w for the white (we assume that only the shell and the white will interact with the vinegar). The mass of the white is the sum of the mass of water in the white (m_w) and the mass of proteins (m_p).

6.9.4 Solving

6.9.4.1 Looking for a Solving Strategy

About the question that we decided to study, different mechanisms can be considered:

1. the production of gas at the surface of eggs;

2. the reaction of acetic acid, and its transformation into acetate;

3. accordingly, the variation of the pH;

4. the expansion of eggs, after the shell is dissolved.

For the theoretical of these phenomena, the knowledge to be used for the calculation is straightforward.

6.9.4.2 Implementing the Strategy

Let us examine these four phenomena, introducing calculations.

1. First, bubbles appear at the surface. How much gas is produced?

   Bubbles appear because acetic acid reacts with calcium carbonate, according to the equation that we shall try establish.

   First, calcium carbonate has the chemical formula $CaCO_3$. And acetic acid has the formula CH_3COOH. I met students that do not remember the formula for acetic acid, but why would they not, in such a case, write simply AcH? They could do a lot with this.

   Anyway, we have to begin with these two compounds:

$$CaCO_3 + CH_3COOH \rightarrow ?$$

Among the reaction products, we know that there are bubbles of carbon dioxide (if you want to check that the bubbles are indeed carbon dioxide, you can direct the bubbles toward a solution of calcium hydroxide, and it will get turbid, through the precipitation of calcium carbonate):

$$CaCO_3 + CH_3COOH \rightarrow CO_2 +?$$

Of course, this occurs because acetic acid is an acid, releasing the acetate ion, as well as a calcium ion:

$$CaCO_3 + CH_3COOH \rightarrow Ca^{2+} + CH_3COO^- + CO_2$$

Now, we have to "equilibrate" the reaction. The hydrogen ion is released by acetic acid because of interactions with water:

$$CaCO_3 + CH_3COOH \rightarrow Ca^{2+} + CH_3COO^- + H_2O + CO_2$$

But now, you see that because calcium is divalent, we need two acetates to make the solution electrically neutral:

$$CaCO_3 + 2CH_3COOH \rightarrow Ca^{2+} + 2CH_3COO^- + H_2O + CO_2$$

And that's it!

Let us now move to the calculation of the volume of gas produced.

In the above equation, for 1 mole of $CaCO_3$ (molar mass $M_{CaCO_3} = 40 + 12 + (3\times16)$ 100 g), 1 mole of CO_2 (24 L at 20°C) is produced.

Assuming that the shell is entirely made of calcium carbonate, one can determine the number of moles of calcium carbonate:

$$n = \frac{M_s}{M_{CaCO_3}}.$$

Because the volume of one mole of gas (carbon dioxide) is 24 L at 20°C under ordinary pressure, the total volume of gas being produced is simply:

$$\frac{M_s}{M_{CaCO_3}} \cdot 24.$$

2. If the egg is simply covered by vinegar, is there enough acetic acid for dissolving the entire shell?

Above, we calculated the number of moles of calcium carbonate:

$$n = \frac{M_s}{M_{CaCO_3}}$$

In the chemical equation of the reaction, two moles of acetic acid are needed for dissolving one mole of calcium carbonate, so that $2\,n$ moles of acetic acid are needed for the dissolution of the entire shell.

Thus the mass of needed acetic acid is the product of the number of moles by the molar mass of acetic acid:

$$m_{CH_3COOH} = 2 \cdot \frac{M_s}{M_{CaCO_3}} \cdot M_{CH_3COOH}$$

Vinegar being a solution with a mass proportion p of acetic acid, the mass of needed vinegar is:

$$m_{vinegar} = \frac{m_{CH_3COOH}}{p} = \frac{1}{p} \cdot 2 \cdot \frac{M_s}{M_{CaCO_3}} \cdot M_{CH_3COOH^3}.$$

3. About the pH, we have two cases to consider:

- the initial pH.

- the final pH.

For the first one, it is assumed to be the one of a solution of acetic acid at p % in mass.

For its determination, let us begin by considering the dissolution of acetic acid in water. This can be described by the chemical equation:

$$CH_3COOH + H_2O \rightleftharpoons CH_3COO^- + H_3O^+$$

Here, the double arrow indicates an equilibrium). In practice, the solution contains non-dissociated (electrically neutral) acetic acid CH_3COOH, acetate ions CH_3COO^-, water molecules, hydrated protons H_3O^+, and hydroxide ions OH^-, because of the second equilibrium, for water:

$$2H_2O \rightleftharpoons OH^- + H_3O^+$$

The question is now to calculate the concentration of each of the chemical species, in particular the concentration of H_3O^+ ions, in order to determine the pH (one remembers that pH means $-\log[10]([H_3O^+])$.

As said before, of course one could remember the formula for this particular case, i.e., pH $= \frac{1}{2}(pK_A + pc)$, where c is the concentration of acid and this particular p is short for $-\log[10]()$, but not all students remember it. As said before, chemical education often confused chemistry and the calculation needed for it: many students were lost in making the right approximations for solving an equation that they could eventually establish.

Let us go slowly. First, let us draw the formula of acetic acid, using a chemistry software (Figure 6.21).

Acetic acid includes a carboxylic acid group -COOH, and its "acidity" is characterized by its pK_a (Lide, 2005):

$$pK_a = 4.756$$

Now, the question studied is: if some acetic acid is put in water, what is the pH

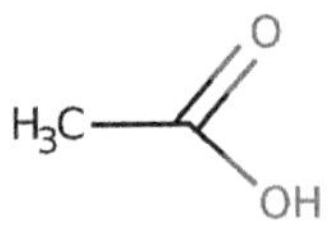

FIGURE 6.21 *Acetic acid drawn using the software* MarvinSketch *17.21. Such software is very helpful for beginners: if you forget everything about acetic acid, you simply type the name, and you get the formula as well as a lot of properties, such as melting temperature,* LogP, *NMR spectrum, ... and* pKa.

of the solution? Again, we shall use here the "backbone for calculation" that we use systematically throughout this book.

We begin by observing that pH is a quantity, but contrary to others, it is written in roman, and not in italics: certainly, there is something behind this exception, and we could look for the reason in Jensen (2004).

For this calculation, we shall again use *Maple* (or another symbolic computation program). It is always good to repeat the question: what is the pH of a solution made by the addition of a known quantity of acetic acid in a known quantity of water?

The data is the pK_a for acetic acid, and the pK_e for water. Now, the qualitative description is the chemical analysis of phenomena without trying to get quantitative. For short, let us name acetic acid AH, so that in water it reacts according to:

$$AH + H_2O \rightleftharpoons Ac^- + H_3O^+$$

For water dissociation:

$$2H_2O \rightleftharpoons H_3O^+ + OH^-.$$

In order to become quantitative, one has to:

- give the quantity of acetic acid and the quantity of water.

- write the equations for matter conservation.

- write the equations for electricity conservation.

- write the equations for energy conservation.

As explained before, we use short variables (a, b, c, d) instead of the full conventional names ([AH], [A$^-$], [H$_3$O$^+$], [OH$^-$]):

$a = [AH]$
$b = [A^-]$
$c = [H_3O^+]$
$d = [OH^-]$

1. For the conservation of matter, let m be the mass of acetic acid (molar mass M, introduced in the volume V of water. Then the concentration C is:

$$C = \frac{\left(\frac{m}{M}\right)}{V}$$

And the conservation equation is:

$$gc(): restart:$$

$$eq1 := \frac{\left(\frac{m}{M}\right)}{V} = a + b$$

$$eq1 := \frac{m}{MV} = a + b$$

2. The "conservation of electricity" means that the solution is neutral, *i.e.*, has many positive charges as negative charges:

$$eq2 := b + d = c$$

$$eq2 := b + d = c$$

3. For the "conservation of energy", this means expressing K_a and K_e:

$$eq3 := Ka = \frac{b \cdot c}{a}$$

$$eq3 := Ka = \frac{bc}{a}$$

$$eq4 := Ke = c \cdot d$$

$$eq4 := Ke = cd$$

Now, we observe that we have 4 equations and 4 unknowns, so that a solution can most likely be found. Anyway, we show the resolution in details, as would a novice user of *Maple*. With brute resolution:

$$solve(\{eq1, eq2, eq3, eq4\}, \{a, b, c, d\})$$

$$\Big\{ a = (-MV\,RootOf(MV\,\_Z^3 + KaMV\,\_Z^2 + (-KeMV - Kam)\_Z - KaKeMV)^2$$

$$+ KeMV + mRootOf(MV\,\_Z^3 + KaMV\,\_Z^2 + (-KeMV - Kam)\_Z - KaKeMV))/$$

$$(RootOf(MV\,\_Z^3 + KaMV\,\_Z^2 + (-KeMV - Kam)\_Z - KaKeMV)MV),$$

$$b = -(-RootOf(MV\,\_Z^3 + KaMV\,\_Z^2 + (-KeMV - Kam)\_Z - KaKeMV)^2 + Ke)/$$

$$(RootOf(MV\,\_Z^3 + KaMV\,\_Z^2 + (-KeMV - Kam)\_Z - KaKeMV)),$$

$$c = RootOf(MV\,\_Z^3 + KaMV\,\_Z^2 + (-KeMV - Kam)\_Z - KaKeMV),$$

$$d = \frac{Ke}{RootOf(MV\,\_Z^3 + KaMV\,\_Z^2 + (-KeMV - Kam)\_Z - KaKeMV)} \Big\}$$

You observe that *Maple* gives many solutions, because it calculates with complex numbers. And using real numbers would not be enough:

$$assume(0 < a): \ assume(0 < b): \ assume(0 < c): \ assume(0 < d):$$

use *RealDomain* **in** $solve(\{eq1, eq2, eq3, eq4\}, \{a, b, c, d\})$ **end use**

$$\left\{ a \sim= (-MV \, RootOf(MV \_Z^3 + KaMV \_Z^2 + (-KeMV - Kam)\_Z - KaKeMV)^2 \right.$$

$$+ KeMV + mRootOf(MV \_Z^3 + KaMV \_Z^2 + (-KeMV - Kam)\_Z - KaKeMV))/$$
$$(RootOf(MV \_Z^3 + KaMV \_Z^2 + (-KeMV - Kam)\_Z - KaKeMV)MV),$$

$$b \sim= -(-RootOf(MV \_Z^3 + KaMV \_Z^2 + (-KeMV - Kam)\_Z - KaKeMV)^2 + Ke)/$$
$$(RootOf(MV \_Z^3 + KaMV \_Z^2 + (-KeMV - Kam)\_Z - KaKeMV)),$$

$$c \sim= RootOf(MV \_Z^3 + KaMV \_Z^2 + (-KeMV - Kam)\_Z - KaKeMV),$$

$$\left. d \sim= \frac{Ke}{RootOf(MV \_Z^3 + KaMV \_Z^2 + (-KeMV - Kam)\_Z - KaKeMV)} \right\}$$

Another possibility is to give immediately numerical data, in order to solve numerically:

$$solve(-\log[10](Ka) = 4.765, Ka)$$

$$1.72 \times 10^{-5}$$

$$M := 60:$$
$$Ke := 1e - 14:$$
$$m := 5:$$
$$Ka := 0.00001717908387:$$
$$V := 1:$$

Using these values, one solves the system in real numbers:

use *RealDomain* **in** $solve(\{eq1, eq2, eq3, eq4\}, \{a, b, c, d\})$ **end use**

$\{a \sim= 0.007945204908, b \sim= -0.0003694473325, c \sim= -0.0003694473596,$
$d \sim= -2.706745560 \, 10^{-11}\}, \{a \sim= -5.821032177 \, 10^{-10}, b \sim= 0.007575758158,$
$c \sim= -1.319999898 \, 10^{-12}, d \sim= -0.007575758159\}, \{a \sim= 0.007223489327,$
$b \sim= 0.0003522682486, c \sim= 0.0003522682770, d \sim= 2.838745539 \, 10^{-11}\}$

The last solution contains only real and positive numbers: it is the one that we are looking for:

$$\{a\sim = 0.007223489327, b\sim = 0.0003522682486, c\sim = 0.0003522682770,$$
$$d\sim = 2.838745539\,10^{-11}\}$$

From it, the pH can be calculated:

$$-\log[10](0.007223489327)$$

$$2.14 \times 10^0$$

For the validation, one could use the approximate equations from the textbooks:

$$pH = \frac{1}{2} \cdot (pKa + pc)$$

Now, for the final pH, the same strategy holds.

4. When the quantity of vinegar is enough, the shell is entirely dissolved, and then the egg swells. Why?

Because of osmosis. This should be an opportunity to correct a mistake that many people make, describing osmosis by saying that "water moves from the more concentrated solution toward the less concentrated solution". This is not true: at any time, even when an equilibrium is reached, water molecules move in both directions through the egg membrane. However, water molecules are more attracted by ions and hydrophilic molecules (Marbach and Bocquet, 2019).

Here, the two solutions exchanging through the membrane of the egg are:

- the egg white, *i.e.*, a 10% solution of proteins;
- the outside solution, with calcium acetate and possibly acetic acid.

The issue, for this exercise, is to guess what can be calculated, and indeed the swelling is not an easy question because if we can easily determine the osmotic pressure simply by applying the classic formula, the resistance of the egg membrane is not known.

The osmotic pressure Π is given by the Van 't Hoff formula (McQuarrie and Simon, 1997):

$$\Pi = cRT$$

where c is the molar concentration of solute, R is the ideal gas constant, T the absolute temperature (in kelvins). This formula applies when the solute concentration is sufficiently low that the solution can be treated as an ideal solution.

But this applies only for one solute against pure solvent. Here, one has to make other assumptions, such as:

- the solutions are ideal;
- acetic acid and proteins of the egg white do not cross the membrane (which is not true, because after some days or weeks, the egg coagulates due to acid migration inside the egg white).

One could calculate the difference between two osmotic pressures.

6.9.5 Expressing the Results

6.9.5.1 *The Formal Result Found is Written Down Again.*

1. Concerning the first discussion, we can decide to calculate the volume of bubbles formed. In L, it is:

$$v = \frac{M_s}{M_{CaCO_3}} \cdot 24$$

2. About the question of acetic acid dissolving the shell, the mass of vinegar needed is:

$$m_{vinegar} = \frac{1}{p} \cdot 2 \cdot \frac{M_s}{M_{CaCO_3}} \cdot M_{CH_3COOH}$$

3. This was done above because we needed to turn digital in order to solve the system of equations.

4. Now, about the question used as a title of this exercise, we have to admit that in ignoring the resistance of the semi permeable membrane around the egg white, we cannot calculate the volume increase. Nonetheless, we can compare the osmotic pressures due to proteins inside (concentration c_p), and acetic acid outside (concentration c_a).

For the egg white:
$$\Pi_p = c_p \cdot R \cdot T$$

For acetic acid:
$$\Pi_a = c_a \cdot R \cdot T$$

The difference can be expressed as a function of the height of water in a tube:

$$P = \varrho \cdot g \cdot h$$

where ϱ is the density of water, g the acceleration of gravity, and h the height of water:

$h = \frac{1}{\varrho g} \cdot (c_p \cdot R \cdot T - c_a \cdot R \cdot T).$

6.9.5.2 *Finding Digital Data*

We already found the needed data in previous exercises.

6.9.5.3 Introduction of Data in the Formal Solution

1. Assuming a mass of shell of 5 g:

$$subs \left(Ms = 5, MCaCO_3 = 100, v = \frac{Ms}{MCaCO_3} \cdot 24 \right)$$

$$v = 1.20$$

This result is expressed in L.

2.

$$subs \Bigg(p = 0.05, r = 0.1, M = 60, MCaCO_3 = 100, MCH_3COOH = 60,$$

$$mvinegar = \frac{1}{p} \cdot 2 \cdot \frac{r \cdot M}{MCaCO_3} \cdot MCH_3COOH \Bigg)$$

$$mvinegar = 1.44 \times 10^2$$

This result is expressed in g.

3. See above.

4. As always, we use SI units, but we first have to determine the concentrations of acetic acid and of proteins.

For acetic acid, we assume a 6% solution. This would make about 60 g of acetic acid; and as the molar mass of acetic acid is about 60 g, the concentration is about 1 mol/L.

For proteins, we assume that they are like ovalbumin, *i.e.*, with a molar mass of 45,000 g. With 3 g (10%) of proteins in egg white, this makes a concentration (in mol/L) of:

$$\frac{3}{45000} = 6.67 \times 10^{-5}.$$

With these values:

$$subs \Bigg(R = 8.32, \varrho = 1000, T = 300, g = 9.8, ca = 1, Ma = 60, cp = 6.67e - 5,$$

$$Mp = 45000, h = \frac{1}{\varrho g} \cdot (ca \cdot R \cdot T - cp \cdot R \cdot T) \Bigg)$$

$$h = 0.2546768894$$

This result is expressed in m. Because it is a small value, it means that the membrane of egg is weak.

6.9.5.4 Discussion

The main discussion could be about osmosis and the resistance of the egg membrane, and this discussion would be useful because there are many misconceptions about eggs. For example, many students think that the yolk is at the center of the egg because of the chalazes. Indeed, these membranes are very soft, and they don't prevent the yolk from floating in the egg white due to lower density (there are fats in it).

6.9.6 Conclusions and Perspectives

It is useful here to discuss more in depth the expression of the results. Often, we simply gave a pre-formatted expression given by the software, but more work should be done in order to be sure of the final expression with only significant figures. Certainly, we have already done such a thing, but students find it useful to have more examples.

Two main ideas have to be given:

1. uncertainties are quantities of the second order of magnitude, and one can calculate them only after the first order is made; *i.e.*, when the quantities themselves are determined;

2. next comes the question of estimating the uncertainties on the various parameters, and this can be the precision of a measurement tool, or the number of figures used in a numerical value, or a standard deviation, for example. One frequent experimental situation is when using a precision scale, say with a possibility to determine the 1/100 of mg. Imagine that because of experimental conditions it is not possible to get the same value three times; then, the standard value for the three measurements is to be used, but if the three measurements are the same, the standard deviation will be nil, and the precision of the balance is to be used.

Let us now put that in action with the above equation:

$$v = \frac{M_s}{M_{CaCO_3}} \cdot 24$$

Here, it would be even better to express it differently, because the number 24 is indeed known more precisely, and should be replaced by the exact volume of one mole of ideal gas:

$$v = \frac{M_s}{M_{CaCO_3}} \cdot \overline{V}.$$

where $\overline{V}$ is the volume of one mole of ideal gas, equal to:

$$\overline{V} = \frac{1 \cdot R \cdot T}{P}$$

Here, we have to use the more precise values:

$$R = 8.3144621 \text{ J} \cdot \text{mol}^{-1} \cdot \text{K}^{-1}$$

$$T = 273.15 \text{ K}$$

$$P = 101325 \text{ Pa.}$$

So that finally, the expression of v is:

$$v = \frac{M_s}{M_{CaCO_3}} \cdot \frac{R \cdot T}{P}$$

Using the guide from the BIPM (2008), we know that the uncertainty on v is the square root of a sum of products, each product being the square of the partial derivative of v relative to one of its variable, by the square of the uncertainty on this variable:

$$\Delta f = \sqrt{\sum_{i=1}^{n} \left(\left(\frac{\partial}{\partial x_i} f \right)^2 \cdot \Delta x_i^2 \right)}$$

Using software, this calculation is easy, even if it is cumbersome. Let us begin by calculating the partial differentials:

$$\frac{\partial}{\partial Ms} \left(\frac{Ms}{MCaCO_3} \cdot \frac{R \cdot T}{P} \right)$$

$$\frac{RT}{MCaCO_3 P}$$

$$\frac{\partial}{\partial MCaCO_3} \left(\frac{Ms}{MCaCO_3} \cdot \frac{R \cdot T}{P} \right)$$

$$-\frac{MsRT}{MCaCO_3{}^2 P}$$

$$\frac{\partial}{\partial R} \left(\frac{Ms}{MCaCO_3} \cdot \frac{R \cdot T}{P} \right)$$

$$\frac{MsT}{MCaCO_3 P}$$

$$\frac{\partial}{\partial T} \left(\frac{Ms}{MCaCO_3} \cdot \frac{R \cdot T}{P} \right)$$

$$\frac{MsR}{MCaCO_3 P}$$

$$\frac{\partial}{\partial P} \left(\frac{Ms}{MCaCO_3} \cdot \frac{R \cdot T}{P} \right)$$

$$-\frac{MsRT}{MCaCO_3 P^2}$$

Of course, one could do that easily, but the use of software is a way to avoid mistakes and/or to validate the calculations that we could do by ourselves.

The second step is now to express the uncertainty on v:

$$\Delta v = \left(\left(\frac{RT}{MCaCO_3 P} \right)^2 \cdot \Delta Ms^2 + \left(-\frac{MsRT}{MCaCO_3{}^2 P} \right)^2 \cdot \Delta MCaCO_3{}^2 \right.$$
$$\left. + \left(\frac{MsT}{MCaCO_3 P} \right)^2 \cdot \Delta R^2 + \left(\frac{MsR}{MCaCO_3 P} \right)^2 \cdot \Delta T^2 + \left(-\frac{MsRT}{MCaCO_3 P^2} \right)^2 \cdot \Delta P^2 \right)^{\frac{1}{2}}$$

Here, we know the values for the variable, but we have to estimate the uncertainties for each of them:

- for ΔMs, we can estimate it to 1 g (The Poultry Site, 2008).

- for $MCaCO_3$, we used 100 g, but a more precise value would be 100.0869, which makes a difference $\Delta MCaCO_3$ of about 0.09.

- as well, we have ΔR estimated at $8.32 - 8.3144621 = 0.0055379$.

- for ΔT, as well, we use 0.15 K.

- and for ΔP, we use 10^3 Pa.

Now the calculation is easy, and the value found is 0.0002244275882, so that our two displayed digits were safe. Another question is to know if we used all the significant figures that we could but remember that we used the ideal gas model, and that deviation from this equation has to be taken into consideration as well. Certainly, we have to be cautious when we are using orders of magnitudes, and the lesson would be: do not provide too many digits!

6.10 HOW TO MAKE A COCKTAIL WITH MORE THAN FOUR LAYERS

6.10.1 An Experiment to Understand the Question

6.10.1.1 First, the General Idea

Previously, in Section 6.7, we briefly discussed the cocktail called Irish Coffee. Of course there are many different ways of making it, but I personally like the one with distinct layers, that you can make as follows:

1. prepare a syrup from sugar and water;

2. prepare an espresso coffee;

3. prepare a shot of whiskey;

4. whip cream;

5. for making the layers, use the idea that a biphasic system will be more stable if the denser liquid is at the bottom, and the less dense liquid at the top; as syrup can have a high density, cool it and place it first ; then, as coffee has almost the density of water, add it over the syrup while hot; now, pour whiskey over the coffee, which is very convenient because the density of ethanol is even less than the density of oil—the latter being of course less than the density of water; and add whipped cream on top (it will float because it is made of oil and air, even less dense than whiskey).

You need advice to make whipped cream? It is simply obtained in this way:

- put a vessel and a whisk in the deep freeze for some minutes;

- store cream in the fridge;

- put some cream in the vessel;

- whip the cream for some minutes: you will get a foamy system that is "whipped cream", or "chantilly cream" if you add sugar.

6.10.1.2 Questions of Methods

If you conduct the experiment proposed in the previous paragraph, you will observe that one has to be very careful in pouring one less dense liquid over a denser liquid, such as whiskey on coffee: if care is not taken, the layers mix. In order to avoid it, bartenders pour slowly the less dense liquid over the convexity of a spoon turned upside down in a glass that is tilted, so that the liquid has only a small distance to flow. Carry out your tests but do not forget that one can also use different tools than the ones that bartenders have at their disposal: for example, instead of pouring from cups or pans, you can put the liquids in pipettes from which you control the flow rate more easily.

6.10.1.3 Where the Beauty Lies

Questions of density have many applications, and one should be aware that air has a lower density than ethanol, ethanol has a lower density than oil, and oil has a lower density than water (Lide, 2005).

This can be demonstrated with a test tube in which you put about 2 cm^3 of water, then the same amount of ordinary cooking oil on top, pouring slowly, and, if possible, making the oil flow along a glass rod that contacts with the opening of the oil bottle and one edge of the test tube. When this is done, you add ethanol on oil, using the same kind of process, and you end up with three layers: from bottom to top, water, oil and ethanol.

Now, let us slowly add water in ethanol: the density of the upper ethanol+water mixture increases from the one of pure ethanol toward the one that can become more than the density of oil. When there is an exact match, the ethanol+water mixture begins going down into the oil, and you will even have the possibility of seeing a drop of a liquid in the middle of another.

This is how some fashionable lamps are produced, the bulb being at the bottom of the system: a drop of liquid is heated whose density is slightly higher than the one of the general liquid, so that the drops being less dense, moves upward, cools on arriving toward the top of the system, and goes down again, and so on.

Could not we turn this into an edible system?

6.10.2 The Question

Until now we produced systems with only four layers, but the last experiment showed how one can adjust densities in order to have continuous densities, so that more layers are than three layers are possible. And indeed, the initial question "How to make a cocktail with more than four layers?" does not state whether or not the systems have to reach the thermodynamical equilibrium: after all, cocktails are not created for infinite stability and they will be consumed sooner or later.

6.10.3 Analysis of the Question

6.10.3.1 The Data is Introduced

We need the density values for the various compounds: air, water, oil, and ethanol.

For air, we could use the fact that 1 mole has a mass of about 29 g (remember the density of gases, equal to $M/29$, where M is the molar mass of a particular pure gas); as the volume would be 24 L at 20°C, one can calculate a density (in kg/m³) of:

$$\frac{29e-3}{24e-3} = 1.21$$

Indeed, in tables, the value found is 1.2 kg/m³ at sea level, at 15°C (Lide, 2005).

For water: we know the value of 1000 kg/m³, but there are variations with temperature. In particular, we are interested in ice cold water (density 999.8426 kg/m³), at 20°C (998.2063 kg/m³) or at the maximum that one can drink (977.63 kg/m³).

For oil, the value chosen could be chosen as 900 kg/m³.

For ethanol, the density is 789 kg/m³.

In this list sugar is missing, and one could first calculate a very crude estimation of density, assuming that the addition of sucrose in water does not change the volume; with a saturating quantity (1.9 kg/L), it would make a density of 2900 kg/m³. Of course, this is wrong because there is indeed a volume increase.

In the past, special units were introduced for measuring the concentration via density. One is the degree Baumé, and another is the Brix, equal to Baumé divided by 0.55.

In the kitchen this is measured by a densimeter, and there are corresponding tables, between the mass of sugar and the density.

6.10.3.2 Qualitative Model

Let us begin with a picture (Figure 6.22). We know that we want to make layers. And some stability will be reached when the layers are organized by order of decreasing density, from bottom to top.

FIGURE 6.22 *Layers for a cocktail. In order to make them more simply, the denser layers are at the bottom, as well as the coldest.*

6.10.3.3 Quantitative Model

Remember that we now have to make the layers, along with our calculations. And also we have to observe that the solutes responsible for color, taste, and odor are present in beverages in low concentration so that we can discuss the question of organizing layers with water, sucrose, oil, ethanol, and air without taking the issue of flavor into account.

Solutions of sucrose in water (syrups) will be put at the bottom because they have the highest density, with, on top, pure water, then emulsions, then a mixture of water and ethanol, then an emulsion made of water, oil, and ethanol, and finally foams.

Certainly, no difficult calculation is needed for the determination of densities, as they can be assumed as barycenters of the various components.

We already discussed the question of sucrose syrup, whose density is much more than for coffee. And we previously discussed the process of rate zonal centrifugation, in which various layers of sucrose syrups are organized. Often, instead of continuous gradients, layers are added, with steps due to 10% increase in solute concentration.

This means that about ten layers can be made from the most concentrated syrup to water. The upper layer would have a density of 1000 kg/m^3. What can we put over it? An emulsion, made of oil droplets dispersed into water, is less dense than water: you can calculate the density of the emulsion ϱ_e from the density of water ϱ_w and the density of oil ϱ_o as:

$$\varrho_e = p\varrho_w + (1-p)\varrho_o$$

with p the proportion in water. Let us observe that p can be as low as 0.05 (5%), but also that various emulsions would not mix easily, because their viscosity can be high. Here, if you want to calculate this viscosity ν, you can apply -it is a brutal application- the Einstein formula (This vo Kientza, 2021):

$$\nu = \nu_0 \cdot (1 + 2.5\Phi)$$

with ν_0 the viscosity of water, and Φ the proportion of the dispersed phase (the volume fraction of the dispersed particles). Mind that this equation holds only for dilute systems made of hard spheres: it would be silly to consider that it applies rigorously, but it can be used as a start.

For foams, the same kind of calculations can be done but with the density of air instead of the density of oil.

For systems containing ethanol and water, one should be aware that the curve of the density as a function of the proportion of ethanol is not straightforward, but indeed a line could also be used as a first approximation. And this assumption could also be made for an emulsion of oil into a solution of ethanol in water.

Finally, no rule forbids making a jellified layer, which means that we could have produced an infinite number of gels one over the other, at the bottom of the glass. And also because foams don't mix (consider whipped egg white, and whipped cream, for example), layers of various foams in large number could be deposited over the liquid layer arrangement. You see: it is not difficult to make more than the three layers of Irish Coffee... and this is the reason why "molecular mixology" was such a success in bars around the world!

6.10.4 Solving

6.10.4.1 Looking for a Solving Strategy

For this chapter the previous analysis gave the solution, and only a choice of flavor has to be made: this is a question of (culinary) art, not technique or science.

6.10.4.2 Implementing the Strategy

As seen, the possibilities are infinite. And it will be enough to observe that a first achievement was demonstrated in 2005, during a training session in "molecular mixology" that was given to the 20 bartenders from around the world, at the Ritz hotel in Paris. This nine-layer cocktail was shown as an introduction to the day: it was performed from my proposal, under my direction, by the famous bartender of the hotel, and the name that I gave it was "Welcome Coffee" (This vo Kientza and Gagnaire, 2021d).

The first layer from the bottom is a layer of jellied coffee. It is simple to make: you dissolve gelatin in coffee, and then pour a layer on the bottom of the glass.

Because coffee pairs well with chocolate, another layer, this one liquid, is poured on top of the first: it is a layer of dense and cold chocolate.

On top of this, a layer of the same chocolate, this one hot, is carefully poured: due to the difference in temperature, the layer of hot chocolate does not mix with the layer of cold chocolate... and the lower jelly layer is protected by the cold chocolate.

In order to place a layer on top of the upper layer of hot chocolate, play with temperatures will no longer be sufficient. A liquid with a density lower than that of water must be used, which makes up the hot chocolate. For example, coffee oil, obtained by macerating oil with coffee powder, is used and emulsified in coffee with added gelatin. The emulsion will have a density almost equal to that of the oil so it will be placed on the hot chocolate without mixing with it.

Even less dense? Vodka, or whiskey! Slowly pour it over the coffee emulsion: it will remain there without mixing.

Still less dense, a gas is required. Or, more exactly, a foam: I suggested dissolving gelatin in sweet coffee, then whipping vigorously for a long time. A rich coffee foam forms. We will pour it delicately on the vodka.

On top of this, I proposed to finally put a "wind crystal", like an aerated meringue. A few roasted almonds, with a little bit of a mixture of tartaric acid and sodium bicarbonate, to foam at the time of consumption. The effect is guaranteed.

Let us count the layers: (1) jelly; (2) cold chocolate; (3) hot chocolate; (4) emulsion; (5) vodka; (6) gelled mousse; (7) solid mousse; (8) almonds; (9) effervescent powder.

Nine layers for a cocktail: that's a nice "molecular mixology" achievement, isn't it?

It is only given as an example. The consideration of densities makes it possible to transpose this cocktail with tailor-made tastes. Here, we chose flavours that pair well with coffee flavour, but we could be more general, in the kitchen: do the same kind of culinary production with broths, strawberry juice, etc.

6.10.5 Expressing the Results

It is easy to understand that here, no big result is needed; contrary to that in other chapters: the proof is in the drinking!

6.10.6 Conclusions and Perspectives

Two last thoughts:

1. If you use some liquids such as Schweppes, you can obtain a remarkable effect when you illuminate the cocktail with ultraviolet radiation because quinine (IUPAC name: (1R)-(6-methoxyquinolin-4-yl)((2S)-5-vinylquinuclidin-2-yl)-methanol) (Figure 6.23) is fluorescent (Chen, 1967).

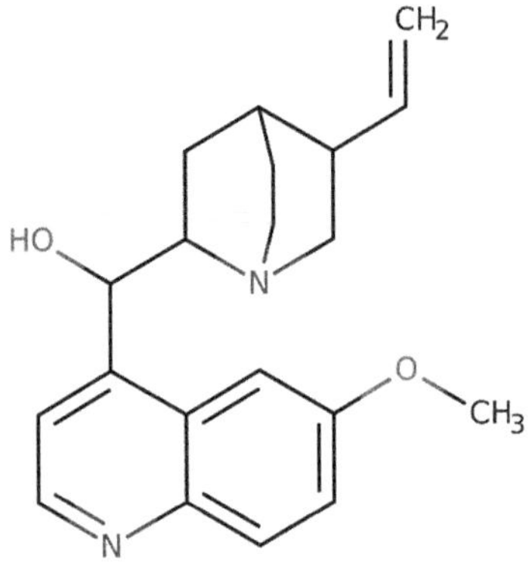

FIGURE 6.23 *Quinine.*

2. Also, all the previous discussion considered horizontal layers... but why not vertical ones? Imagine that you tilt the glass before making your cocktail and that you created a jellified layer. Indeed, this means "thinking out of the box", and it is very important for technology.

REFERENCES

AOAC. 2000. *AOAC official method 934.01. Loss on Drying (Moisture) at 95–100° C for Feeds. Dry matter on oven drying at 95–100° C for feeds*, https://www.edgeanalytical.com/wp-content/uploads/Food_AOAC-934.01.pdf, last access 2022-01-10.

When you need analytical methods, go to this source because they are well validated. Course, you can devise your own experiment, but then you will have to show that it is right.

Archer DG, Wang P. 1990. *The dielectric constant of water and Debye-Hückel limiting law slopes*, Journal of Physical and Chemical Reference Data, 19, 371.

Atkins PW. 1984. *The second law*, Scientific American Library, New York, USA.

When I was using Atkins' Physical Chemistry, before moving on to McQuarrie's book, I was dreaming of a book that would discuss the question of the second principle of thermodynamics... and Atkins produced this remarkable book! A good complement to the Berkeley course of physics (statistical thermodynamics).

Belitz HD, Grosch W, Schieberle P. 2009. *Food chemistry*, Springer Verlag, Heidelberg, Germany.

Not the first time that this title is cited, in this book, and not the last one.

Bieze TWN, van der Maarel JRC, Leyte JC. 1993. *The intramoleular OH bond length of water in a concentrated poly(ethyleneoxide) solution. An NMR relaxation study*, Chemical Physics Letters, 216(1,2), 56–62.

BIPM. 2008. *Evaluation of measurement data — Guide to the expression of uncertainty in measurement, JCGM.* https://www.google.com/url?sa=t&rct=j&q=&esrc=s&source=web&cd=&cad=rja&uact=8&ved=2ahUKEwjx8t_Ml6D1AhXwgc4BHfskD0oQFnoECAQQAQ&url=https%3A%2F%2Fwww.bipm.org%2Fdocuments%2F20126%2F2071204%2FJCGM_100_2008_E.pdf%2Fcb0ef43f-baa5-11cf-3f85-4dcd86f77bd6&usg=AOvVaw36IcV_QMF8qwNKNE1ZO_mF, last access 2022-01-07.

Certainly, this document is not the most amusing source to read, but it helps prevent us from making bad mistakes.

Cabane B, Vuilleumier R. 2005. *The physics of liquid water,* Comptes Rendus Géoscience, 337(1-2), 159–171.

Here, I can tell you how angry I am! Before finding this good reference, I found online too many pages, in which the information was given with no scientific reference: this is a very bad practice because how can you be sure of the information given? In the end, you need to see a scientific article that provides a good "Material and methods" section, explaining how the determination was made.

More generally: never trust someone who simply gives you an unreferenced piece of data (except perhaps for popular science, which is another story).

Carpenter WB, Clymer M, Smith FG. 1845. *Principles of human physiology,* Lea & Blanchard, London, UK.

Chaplin M. 2020. *Water structure and science,* http://www1.lsbu.ac.uk/water/water_models.html, last access 2022-01-07.

Chemistry-LibreTexts. 2020. *A3: Covalent radii,* https://chem.libretexts.org/Ancillary_Materials/Reference/Reference_Tables/Atomic_and_Molecular_Properties/A3%3A_Covalent_Radii, last access 2022-01-07.

Chemistry Stack Exchange. 2021. *Why is anhydrous copper(II) sulfate white while the pentahydrate is blue, even though both have one unpaired electron?,* https://chemistry.stackexchange.com/questions/71980/why-is-anhydrous-copperii-sulfate-white-while-the-pentahydrate-is-blue-even-t, last access 2022-01-10.

Engineering ToolBox. 2009. *Volumetric expansion coefficients liquids,* https://www.engineeringtoolbox.com/cubical-expansion-coefficients-d_1262.html, last access 2022-01-07.

I could have given many different references here, but this one is particularly interesting: first, because there is the beginning of a good discussion, leading to other sources, but also because Chemistry Stack Exchange is a good way to explore chemistry.

Chen RF. 1967. Some characteristics of the fluorescence of quinine, *Analytical Biochemistry,* 19, 374–387.

Ciqual. 2020. *French food composition table*, https://ciqual.anses.fr/#, last access 2022-01-10.

ECHA. 2020. *Sulfur, substance inforcard.* https://echa.europa.eu/substance-information/-/substanceinfo/100.028.839.

EFSA. 2008a. *Conclusion on pesticide peer review, peer review of the pesticide risk assessment of the active substance sulfur*, Question No EFSA-Q-2008-293, EFSA Scientific Report 221, 1–70.

EFSA. 2008b. *EFSA opinion on suitable indicators for both the occurrence and toxicity of polycyclic aromatic hydrocarbons (PAHs) in food*, https://www.efsa. europa.eu/en/news/efsa-opinion-suitable-indicators-both-occurrence-and-toxicity-polycyclic.

This source is particularly interesting because it leads to the question: how is it possible that people eat meat grilled on the BBQ, full of benzopyrene, and at the same time say that they want "healthy food"?

EFSA. 2015. *Scientific opinion on the re-evaluation of chlorophylls (E 140(i)) as food additives*, EFSA Journal, 13(5), 4089.

Eugster CH, Märki-Fischer E. 1991. *The chemistry of roses, Angewandte Chemie*, 30, 654–672.

Eurostat. 2020. *Electricity price statistics*, https://ec.europa.eu/eurostat/statistics-explained/index.php/Electricity_price_statistics, last access 2022-01-10.

Factorovich MH, Molinero V, Scherlis DA.2014. *Vapor pressure of water nanodroplets*, Journal of the American Chemical Society, 136, 4508–4514.

Feynman R, Leighton R, Sands M. 1963. *The Feynman lectures on physics*, The California Institute of Technology. On Amazon, I see Basic Books; 50th New Millennium ed. edition (October 4, 2011).

Fisher WD, Cline GB. 1963. *A density gradient for the isolation of metabolically active thymus nuclei*, Biochimica et Biophysica Acta (BBA) - Specialized Section on Nucleic Acids and Related Subjects, 68, 640–642.

Fulcrand H, Morel-Salmi C, Mané C, Poncet-Legrand C, Vernhet A, Cheynier V. 2006. *Tannins: from reactions to complex supramolecular structures, AVSO Proceedings; Advances in tannin and tannin management*, https://www. researchgate.net/profile/Cecile_Morel-Salmi/publication/236015321_Tannins_ from_reactions_to_complex_supramolecular_structures/links/ 5aab8c950f7e9b882671a206/Tannins-from-reactions-to-complex-supramolecular-structures.pdf.

Please be careful: a phenol is a compound whose molecules contain one or more aromatic ring with one or more hydroxy groups. Tannins, in particular, are a special category of phenols, or "phenolic compounds".

Garai J. 2009. *Physical model for vaporization*, Fluid Phase Equilibria 283, 77–89.

Gromacs. 2022. *Gromacs*, https://www.gromacs.org/, last access 2022-01-10.

   It is free! And so exciting! A great "thank you" to the colleagues who developed it.

Harr L, Gallagher JS, Kell GS. 1984. *NBS/NRC Steam tables*, Hemisphere Publishing Corp.

Harris DC. 2007. *Quantitative chemical analysis*, W.H. Freeman and Company, New York, USA.

   If you do chemical analysis, you cannot escape using this book. It is comprehensive, simple, and clear.

Huang Y, Zhang X, Ma Z, Li W, Zhou Y, Zheng W, Sun CQ. 2013. *Size, separation, structural order, and mass density of molecules packing in water and ice*, Scientific Reports, 3 (3005), 1–5. DOI:10.1038/srep03005.

   Here, a comment is needed because this is the first time that I have given a DOI, or Digital Object Identifier. Sometimes, one can hesitate between giving a link or a DOI, but let's observe that sites can change and sometimes disappear, whereas DOIs are here forever.

Hui YH, Sherkat F. *Handbook of food science, technology, and engineering - 4 volume set*, CRC Press, 2005, 9–12.

IUPAC.2019. *Proteins*, compendium of chemical terminology (2nd ed), https://doi.org/10.1351/goldbook, last access 2022-01-10.

Jensen JB. 2004. *The symbol for pH*, Journal of Chemical Education, 81(1), 21.

Kaufman SL, Dorman FD. 2008. Sucrose clusters exhibiting a magic number in dilute aqueous solutions, Langmuir, 24(18), 9979–9982.

Kurti N. 1969. *The physicist in the kitchen*, Proceedings of the Royal Institution (1969), 42/199, 451–67.

Kurti N, This-Benckhard H. 1994. *The kitchen as a lab*, Scientific American, 270(4), 120–123,

Lide DR (ed.). 2005. *CRC handbook of chemistry and physics* (85th ed), CRC Press, Boca Raton FL.

Marbach S, Bocquet L. 2019. Osmosis, from molecular insights to large-scale applications, Chemical Society Reviews, DOI: 10.1039/c8cs00420j.

Maxwell JC.1855. *On faraday's lines of force*, Transactions of the Cambridge Philosophical Society, https://www.scribd.com/doc/39568221/Maxwell-On-Faraday-s-Lines-of-Force, last access 2022-01-10.

Marsh KN (ed). 1987. *Recommended reference materials for the realization of physicochemical properties*, Blackwell Scientific Publications, Oxford.

McQuarrie DA, Simon JD. 1997. *Physical chemistry*, University Science Books, Sausalito, California.

Milgrom LR. 2001. *The colours of life: An introduction to the chemistry of porphyrins and related compounds*, Oxford University Press, Oxford, UK.

NIH. 2020. *Download*, https://imagej.nih.gov/ij/download.html.

NOAA. 2014. *Deep light*, https://oceanexplorer.noaa.gov/explorations/04deepscope/background/deeplight/deeplight.html, last access 2022-01-10.

OIV.2018. *Maximum acceptable limits*. http://www.oiv.int/public/medias/3740/f-code-annexe-limites-maximales-acceptables.pdf, Last access 2018-12-14.

OIV.2021. *SO2 and wine: a review*, https://www.oiv.int/public/medias/7840/oiv-collective-expertise-document-so2-and-wine-a-review.pdf, last access 2022-01-10.

Perrin J. 1913. *Les atomes*. Librairie Félix Alcan, Paris. https://gallica.bnf.fr/ark:/12148/bpt6k373955h.image, last access 2018-11-15.
It is certainly translated into English!

Quideau S. 2022. *Why bother with polyphenols?*, Groupe Polyphenols, https://www.groupepolyphenols.com/the-society/why-bother-with-polyphenols/, last access 2022-01-10.

Reinhart A. 2022. *Statistics done wrong*, https://www.statisticsdonewrong.com/, last access 2022-01-10.

SCAT. 2017. Electrostatic explosion test, https://www.youtube.com/watch?v=wJo9sorZ_W8, last access 2022-01-10.

Sengers JV, Watson TR. 1986. *Improved international formulations for the viscosity and thermal conductivity of water substance*, Journal of Physical and Chemical Reference Data, 15, 1291.

Sharon NB. 1985. *Nomenclature of glycoproteins, glycopeptides and peptidoglycans, IUPAC-ICB Joint Commission on Biochemical Nomenclature (JCBN)*, https://iupac.qmul.ac.uk/misc/glycp.html, last access 2022-01-10.

Sigma-Aldrich. 2018. https://www.sigmaaldrich.com/catalog/search?term=7446-09-5&interface=CAS%20No.&N=0+&mode=partialmax&lang=fr&region=FR&focus=product.

Stein PE, Leslie AGW. 1992. *1OVA*, RCSB PDB, https://www.rcsb.org/structure/1ova, last access 2022-01-10.

Stokes GG. 1851. *On the effect of the internal friction of fluids on the motion of pendulums*, Transactions of the Cambridge Philosophical Society, 9, 8-14, https://babel.hathitrust.org/cgi/pt?id=mdp.39015012112531&view=1up&seq=214, last access 2022-01-10.

Sun Y, Jin H, Sun HH, Sheng L. 2020. *A comprehensive identification of chicken egg white phosphoproteomics based on novel digestion approach*, Food Chemistry, DOI: 10.1021/acs.jafc.0c03174.

Suresh SJ, Naik VM. 2000. *Hydrogen bond thermodynamic properties of water from dielectric constant data*, The Journal of Chemical Physics 113, 9727–9732.

The Poultry Site. 2008. *Maintaining egg shell quality*, https://thepoultrysite.com/articles/maintaining-egg-shell-quality.

This H. 2009. *Histoires chimiques de bouillons et de pot-au-feu*, L'Actualité chimique. 336 (11), 14–16.

This H. 2010. *Cours de gastronomie moléculaire N° 2: les précisions culinaires*, Quae/Belin, Paris.

This H. 2016. *"Maillard products" and "Maillard reactions" are much discussed in food science and technology, but do such products and reactions deserve their name?*, Notes Académiques de l'Académie d'agriculture de France, 3, 1–10.

This H. 2017. *Parlons des chlorophylles, et pas de chlorophylle*, https://hervethis.blogspot.com/2017/12/parlons-des-chlorophylles-et-pas-de.html.

This vo Kientza H. 2021a. *Let us have an egg*, Handbook of molecular gastronomy, CRC Press, Boca Raton, USA, 221–226.

This vo Kientza H. 2021b. *The right words for improving communication in food science, food technology, and between food science and technology and a broader audience*, Handbook of molecular gastronomy, CRC Press, Boca Raton, USA, 626–634.

Valverde J, Vignolle M, This H. 2007. *Quantitative determination of photosynthetic pigments in green beans using thin-layer chromatography and flatbed scanner as densitometer.* Journal of Chemical Education, 84, 1505–1507.

This vo Kientza H. 2021c. *Emulsions. Emulsions and surfactants in the kitchen*, Handbook of molecular gastronomy, CRC Press, Boca Raton, USA, 257–264.

This vo Kientza H, Gagnaire P. 2021d. *Molecular mixology: Welcome coffee, a cocktail with ten layers*, Handbook of Molecular Gastronomy, CRC Press, 827–828.

Tyndall J.1868. *Michael Faraday*, Longmans, Green and Co, London, UK.
Ah, Faraday was such a wonderful man!

USDA. 2008. The use of sulfur for the control of varroa, https://portal.nifa.usda.gov/web/crisprojectpages/0200724-the-use-of-sulfur-for-the-control-of-varroa.html, last access 2022-01-12.

Vargaftik NB *et al.* 1983. *International lubles for the surface tension of water*, Journal of Physical and Chemical Reference Data, 12, 817.

CHAPTER 7

Foams (G/L Systems)

P REVIOUSLY, WE CONSIDERED simple systems with only one phase. Here, we move to foams, which are well defined by the IUPAC (2019):

A dispersion in which a large proportion of gas by volume in the form of gas bubbles, is dispersed in a liquid, solid or gel. The diameter of the bubbles is usually larger than 1 μm, but the thickness of the lamellae between the bubbles is often in the usual colloidal size range. The term froth has been used interchangeably with foam. In particular cases froth may be distinguished from foam by the fact that the former is stabilized by solid particles (as in froth flotation q.v.) and the latter by soluble substances.

In the title of this chapter, we read "G/L", and this means that we consider here "liquid foams" only, with gas structures (G) randomly dispersed in a liquid (L) (This vo Kientza, 2021).

7.1 HOW MANY AIR BUBBLES IN A WHIPPED EGG WHITE?

7.1.1 An Experiment to Understand the Question

7.1.1.1 First, the General Idea

We already consider whipped egg white that you make in the following way:

1. in a vessel, put an egg white;

2. use a whisk and whip, pushing air into the liquid;

3. observe the first bubbles floating at the surface of the (yellowish) liquid;

4. continue whipping, and observe the variation in the size of bubbles;

5. after some time, observe the system at the edge, and estimate the number of layers of bubbles;

6. continue whipping; and when the foam has some "consistency", keeping its form after the whisk has been moved, stop whipping.

DOI: 10.1201/9781003298151-7

7.1.1.2 Questions of Methods

From the IUPAC definition, we understand that whipped egg whites are foams.

Now, let us observe that the experiment above was conducted using a whisk, *i.e.*, a very old instrument (in the past, cooks used wicker whips), but this is not a proof that such tools are efficient. Indeed, it is probably more appropriate to define the goal first (making a foam), and then determine the tools to reach this goal: if the goal is to introduce air bubbles in a liquid, why not use a pump, pushing air? Or siphons with various gases such as nitrous oxide or carbon dioxide? These proposal was one that was made as early as 1980, as a basis of "molecular cooking" (this name was coined only in 1999, in Paris; of course, one should not confuse it with the scientific discipline called "molecular and physical gastronomy"). By the way, these two different techniques can produce foams with different sizes of bubbles.

7.1.1.3 Where the Beauty Lies

This paragraph will bring us back into the kitchen, instead of moving us toward science. Let us observe that the difference between pure water and egg white is mainly the presence of 10% proteins. During whipping, the protein molecules "unfold" partially (denaturation) and make layers at the air/water interface, around all air bubbles.

In chicken egg white, we know 158 proteins, 96 being phosphorylated (Sun *et al.*, 2020), and the main one (45%) being ovalbumin. The reason why many proteins have foaming properties is that they are made of alternating hydrophilic and hydrophobic segments; when proteins unfold, their hydrophobic segments are positioned close to air in foams, or to oil in emulsions, and the hydrophilic parts go into the aqueous solution. Indeed, for egg white foams, it was shown that there are probably two layers of proteins at the air/water interface, one more denatured than the other (Le Floch-Fouéré *et al.*, 2010).

Let us now generalize: if proteins can be helpful in making foams, why use such strange substances as the trendy juice of chickpeas? Certainly, this is a liquid that is derived not from animals, and vegetarians will be happy to know that "pulses" in general are plants full of proteins, and plant proteins are now extracted by the tonne by the food industry: if you add one teaspoon of such powders to water and whip, you will get the same kinds of foams as with egg whites (FAO, 2016). And be mindful that you can use any aqueous solution, not only flavorless water!

Another interesting possibility, now with animal protein, is to use gelatin: imagine that you dissolve it in orange juice, by heating; if you whip, you will get liters of foam... that sets when put in the fridge. I called this system a "würtz" (This vo Kientza, 2021).

7.1.2 The Question

Remember that the question here is: "How many air bubbles in a whipped egg white?". It is a very simple question, and the solution, as we shall now see, seems simple. Indeed, the fact that so many students cannot find the answer demonstrates that this "simplicity" is only apparent. Their difficulties are sometimes the result a lack of a strategy for answering questions, in particular during exams, written or oral.

I propose to distinguish two cases:

- when the student knows the answer to a question that he/she is asked;

- when he/she does not know.

Let us consider the second case. If the student answers "I don't know", it is honest, but the teacher cannot give any mark except the lowest. If a random answer is given, this is silly, and the teacher would give a negative mark if possible. If the student does not know but tres to fool the professor, it is even worse than the second case, because there is a will to fool the professor; perhaps we should even exclude such people from an educational curriculum, because this is dishonest behavior.

A good piece of advice is to discuss out loud the various words of the question, introducing all the scientific knowledge that we can. It is good because:

1. it demonstrates our skills of analysis;

2. it shows the examiner that (even if we do not find the solution to the question) we know something;

3. it moves us towards the solution.

Let us observe that this strategy is very similar to the one shown throughout this book with the "backbone for calculation": there is no reason for not using it, and in particular because discussing the terms of the question leads often to the solution.

In conclusion, a good method is helpful, and it often leads to a result. This is what we propose to examine here.

7.1.3 Analysis of the Question

7.1.3.1 The Data is Introduced

Here, no data is given, and it is a good start to ask oneself if we are looking for an exact number, as obtained using formal calculation, or for an order of magnitude: clearly, the second way is to be chosen now!

About egg whites, even if we do not know their mass, we can assume that it is about half the mass of the egg. And if we do not know the mass of eggs, we can find an order of magnitude: it is not 10 g (a 10 cm height of water in a test tube 1 cm^2 of section), and it is not 1000 g (a pack of milk), so a good assumption is 100 g. However, because such a mass would correspond to a rectangular solid of water with a base of area equal to 10 cm^2, and height of 10 cm, we can see that it would be better to choose half of it and we would therefore find 50 g, which is not far from the correct value (about 60 g).

Now, the egg white (about half the whole egg) is made of proteins and water. Can we guess, as before, the concentration of proteins? Of course, complete ignorance cannot guess anything, but students who had vacations by the sea or who stored meat in a brine could know that the maximum quantity of salt dissolved in water is about 350 g per liter, and about 2 kg for sugar (see the previous chapter). The first is much less viscous than an egg white, and the second is more viscous. Of course, viscosity is not directly proportional to concentration only, but these two values can lead to a guess of the right order of magnitude, again.

7.1.3.2 Qualitative Model

A whipped egg white is obtained by whipping an egg white, *i.e.*, an aqueous solution of 10% proteins with a mass (30 g) of about one half of the whole egg's mass.

Let us first make a picture like in Figure 7.1a.

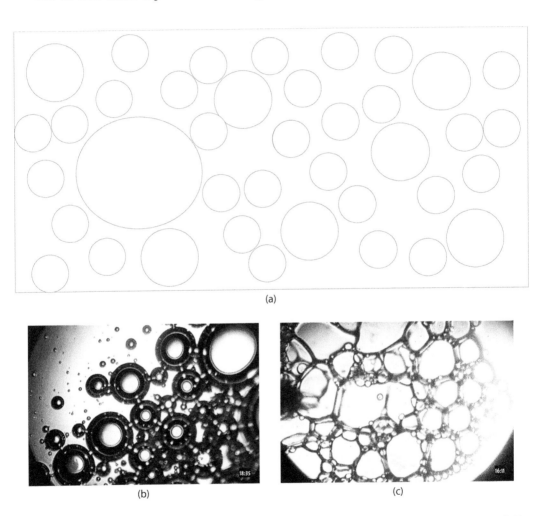

(a)

(b)   (c)

FIGURE 7.1   *A model for a whipped egg white (a). Of course, the students would usefully have to observe a real microscopic picture of a whipped egg white at the beginning of whipping ("wet foam", b) and at the end of whipping (dry foam, c).*

Here, we see that we can add air bubbles into the liquid until they are closely packed, and this should lead us to consider that bubbles could fill up about 75% of the liquid, if all of the same size.

On the other hand, if the bubbles have different sizes, the air fraction could reach a larger proportion.

7.1.3.3 Quantitative Model

Let v be the volume of one egg white, and V the volume after whipping. Let r be the radius of air bubbles, assuming that they are all alike. Let N be their number.

7.1.4 Solving

With N spherical bubbles of air of radius r, the total volume of bubbles is:

$$V_b = N \cdot \left(\frac{4}{3} \pi r^3 \right)$$

The total volume of whipped egg white is the sum of the volume of air and the volume of egg white v:

$$V = V_b + v = N \cdot \left(\frac{4}{3} \pi r^3 \right) + v$$

Here, we look for N:

$$V = N \cdot \left(\frac{4}{3} \pi r^3 \right) + v \xrightarrow{\text{isolate for N}} N = -\frac{3}{4} \frac{-V + v}{\pi r^3}$$

7.1.5 Expressing the Results

7.1.5.1 The Formal Result Found is Written Down Again

In the last section, we found the value of N using the *isolate* command of the formal software, and it appeared with a minus sign that indeed compensates with another one. It seems more convenient to write:

$$N = \frac{3}{4} \frac{V - v}{\pi r^3}$$

7.1.5.2 Finding Digital Data

Having a mass of egg white of 30 g (30×10^{-3} kg), we can find the volume of egg white, assuming a density equal to that of water (30×10^{-6} m^3).

The volume of a whipped egg white is lower than half a liter. Let us decide that it is one-third of a liter (0.3 L, *i.e.*, $0.3\ 10^{-3}$ m^3).

For the volume of an air bubble, one could assume that it is smaller than 1 mm, and indeed, under a microscope we could see air bubbles of various sizes, between 1 mm and 10^{-2} mm; let us use the value 0.1 mm and 0.01 mm.

7.1.5.3 *Introduction of Data in the Formal Solution*

As mentioned, we calculate for two sizes. First, with bubbles of radius 0.1 mm:

$$evalf\left(subs\left(V = 0.3e-3, v = 30e-6, r = 1e-4, -\frac{3}{4}\frac{-V+v}{\pi r^3}\right)\right)$$

6.45×10^7

And then with bubbles of radius 0.01 mm:

$$evalf\left(subs\left(V = 0.3e-3, v = 30e-6, r = 1e-5, -\frac{3}{4}\frac{-V+v}{\pi r^3}\right)\right)$$

6.45×10^{10}

7.1.5.4 *Discussion*

Of course, only the order of magnitude can be kept: the number is between 0.1 billion and 100 billion.

Validation can be done in many ways.

1. For example, let us skip the assumption of a whipped egg white having a volume of 1/3 of a liter, and let us decide instead that air bubbles, spherical and all of the same size, are packed in the liquid egg white. Now, the total volume of air in the liquid is given using the close packing coefficient ($p = 0.74\%$, see Section 7.2). This means that if the volume of the egg white is v, the volume of air is:

$$v_a = \frac{v_l \cdot p}{1-p}$$

   And if the volume of an air bubble is α, then the number of bubbles is:

$$\frac{\frac{v_l \cdot p}{(1-p)}}{a}$$

   This makes:

$$evalf\left(subs\left(p = 0.74, v_l = 30e-6, a = \frac{4}{3} \cdot \pi \cdot (1e-4)^3, \frac{\frac{v_l \cdot p}{(1-p)}}{a}\right)\right)$$

2.04×10^7

   This is the same order of magnitudes as one that we calculated before.

2. Let us now consider that whipping is performed using a whisk with 10 metal wires of 1 mm diameter, and 10 cm of each wire dipping in the egg white at each turn of the whisk. This would make 100 bubbles of 1 mm diameter per thread, *i.e.*, 100×10 bubbles for one turn. If we whip for 3 minutes, with 1 rotation per second, this would make about 200 rotations, *i.e.*, 2×10^5 bubbles.

   With longer whipping (5 minutes) and bubbles divided by 2 at each rotation, the number could increase much more (how much exactly?).

7.1.6   Conclusions and Perspectives

This exercise was primarily a way to examine order of magnitudes, as this was considered of primary importance by Fermi, but also by the French physicist Pierre Gilles de Gennes, who wrote (De Gennes, 1998) that training in orders of magnitude calculation was one way to become "smart" in physics.

Now, a complementary exercise could be to determine the total area of air-liquid interface of the foam in order to understand why edible foams have so much more odor than the odorant liquids from which they are made: in the assumption of an evaporation of odorant compounds proportional to the area of the air/liquid interface, this simple calculation (simply multiply the number of bubbles by the area of their surface) explains the phenomenon well.

7.2   WHEN SUGAR IS ADDED TO A WHIPPED EGG WHITE IN ORDER TO MAKE A MERINGUE, THE SIZE OF THE AIR BUBBLES DECREASES. HOW DOES THE DENSITY OF WHIPPED EGG WHITES CHANGE WITH THE SIZE OF AIR BUBBLES?

7.2.1   An Experiment to Understand the Question

7.2.1.1   First, the General Idea

Here, the starting experiment is simple: let us refer to what is said in the title of the chapter:

1. in a vessel, let us place an egg white;

2. using a whisk, if possible an electrical one in order to get repeatable results, whip until a foam is obtained (not too firm);

3. divide the foam in two, putting the two halves in identical vessels, cleaned in the same way;

4. checking the time with a chronometer, whip one-half until a very firm foam is obtained (measure the time for that);

5. clean the whisk blades;

6. add 100 g of table sugar to the other half and whip for the same time as for the first half;

7. use the tip of a knife to put a small (thin layer) sample of the first foam on the left side of a microscope plate, and a small sample of the second foam on the right;

8. use the lowest magnification of a microscope to observe the two foams: you will be able to observe the size of bubbles and check that adding sugar make smaller bubbles.

7.2.1.2 Questions of Methods

One has to be aware that experiments with foams can be very difficult because the results are very sensitive to "impurities", as there are many in kitchens. In scientific research, the vessels and tools used to make foam research have to be cleaned extensively, with soap first, but then with much care, because soap is made of tensioactive compounds that migrate to the air/water interface and can drastically change the foam structure (Langevin, 2016).

Acids or acetone cause the coagulation of the proteins that would remain on surfaces, so they should not be used. Sodium hydroxide is preferred because it can hydrolyze proteins; however, be sure to prepare the solution at the last moment so that carbonation and possible pollution will be avoided.

7.2.1.3 Where the Beauty Lies

Aqueous foams are made of a gas and a liquid. Certainly, the density of a foam, made of liquid and air, will change if the proportion of the two compounds is changed, and one question is to know if the size of air bubbles can change the density. But we remember from Section 6.10 that the density of a sugar syrup can change as well, and because meringues include sugar, we also have to consider this parameter. And with two control parameters, one can guess that if we intend to make a very light or rather a very heavy foam, so called "optimization methods" can be useful. Do you know about that? Remember, before looking at the equation, that at school, when you look for an extremum (maximum or minimum) of a function f of a real variable x, you calculate the derivative $f'(x) = \frac{d}{dx}(f(x))$ and you solve the equation:

$$\frac{d}{dx}(f(x)) = 0$$

7.2.2 The Question

Some exercises from this book are also useful because they contribute to better observing the phenomena that occur during cooking. Some can be very easy, but after the first easy step is done, more elaborate calculations can be imagined.

Here, the initial question is simple: how does the density of whipped egg white change with the size of air bubbles? However, some students find the answer difficult to arrive at.

7.2.3 Analysis of the Question (Often Questions Are Solved Immediately When They Are Analyzed)

7.2.3.1 The Data is Introduced

No data is needed for this exercise.

7.2.3.2 Qualitative Model

Let us whip an egg white. First, large bubbles (about 2 or 3 cm) appear at the surface, but they are divided by the whisk while new small bubbles appear. Finally, only smaller air bubbles remain throughout the liquid, before becoming too small to be seen with the naked eye (*i.e.*, less than 0.1 mm).

Let us assume that all air bubbles have the same size.

7.2.3.3 Quantitative Model

In Section 7.1, we used the "close packing" coefficient, but we did not explain it. Here, we want to investigate this question slowly, assuming first a little of the background to help with the calculation.

Let us explore the question of bubble packing by making some assumptions that simplify the question: we admit that all bubbles are spherical with the same radius. And because packing spheres in a three-dimensional space is more complex than packing disks in two dimensions, we begin our analysis with disks in a plane: this is a good strategy to study first the simplest question.

How to choose the radius? This has no importance, but some students need to be convinced about this fact, and Figure 7.2 explains it: in this figure we compare the packing of the same "space" (a square of side 1) with either one disk as big as possible, or with four disks of radius two times smaller than the radius of the first disk.

In these two cases, we want to determine the ratio of air over water: it is the ratio of the area of the disks by the area of the blue part outside disks.

If the radius of the large disk in the square is 1, the area is also $\pi.1.1 = \pi$. Now, the area of the blue part is equal to the area of the square, minus the area of the disk, *i.e.*, $4 - \pi$. And the air over water ratio is $\frac{\pi}{4-\pi}$ (about 3.66).

For the second picture, with smaller disks, the radius is $1/2$, so that the area of each square is $\pi (1/2)^2$, and the total area of the white part is equal to $4 \cdot \pi \cdot \left(\frac{1}{2}\right)^2$, that is again π. And the ratio is then the same, which is not surprising because this is the idea of proportionality. In both cases, the space occupied by spheres can be also described by a "density", *i.e.*, the area of spheres compared to the area of the squares, and now we find:

$$d_2 = \frac{\pi}{4} \text{ (about 0.79).}$$

Of course, real foams do not exist in two dimensions, but in a three-dimensional space the analysis would be the same: if you consider a cube of a side equal to 2 (radius of a sphere equal to 1), the volume of the cube would be 8 (I invite you to make the picture by yourself). In it, the biggest sphere that you could have would have a radius equal to 1 and, accordingly, a volume $\frac{4}{3}\pi 1^3$, that is $\frac{4}{3}\pi$. The air/water ratio would be

$$\frac{\frac{4}{3}\pi}{8 - \frac{4}{3}\pi} = \frac{8 \cdot 3}{4\pi}.$$

Now, with the large cube (remember, volume 8) divided into smaller cubes having each a side of 1 only, there would be 8 spheres of radius $1/2$, so that the volume of air and of water would be the same.

This said, we can observe that we did not use the closest packing, but the simplest way to calculate. In two dimensions, a closer packing is obtained when you move a

FIGURE 7.2 *A good way to help students to understand that the packing coefficients holds for any size of bubbles.*

line of disks by one radius on the side so that you can move the row upward, and better fill the space (Figure 7.3).

In three dimensions, if the air bubbles are all of the same size and packed closely, the packing coefficient is about 0.74: this means that the greatest fraction of space occupied by identical spheres is 0.74 times the total volume. Indeed, it was Johannes Kepler who conjectured that the hexagonal close packing was the densest, but this was proved only in 2014, by Thomas Callister Hales (Hales *et al.*, 2017).

Let us come back to our model of meringue now. We consider a meringue, with a liquid and air bubbles all of the same size. The density of the liquid is ϱ_l and the density of air is ϱ_a. The radius of air bubbles is r.

7.2.4 Solving

Here, simply making the model gives the answer.

FIGURE 7.3 *This picture shows that the hexagonal organization makes a closer packing than a square one: if you move one row of disks by one radius on the side, you can move the disks up in order to better fill the space. The calculation of the packing is a good exercise, in two or in three dimensions.*

7.2.5 Conclusions and Perspectives

This exercise being simple, one could try to find other more complex solutions, and, indeed, the assumption of bubbles of all the same size could be dropped.

When there are different sizes, small bubbles can go in the middle of large ones, such as in Figure 7.4a.

Can students calculate the changes in density of whipped egg white in such a case? The answer is yes, and they can work in two or in three dimensions.

In two dimensions, let r_1 be radius of the largest bubbles, and c the compacity (the density is $1-c$).

Imagine that we put a small bubble in the middle of any group of three large bubbles. We see a rectangle triangle, with an angle of 30 ° (Figure 7.4b).

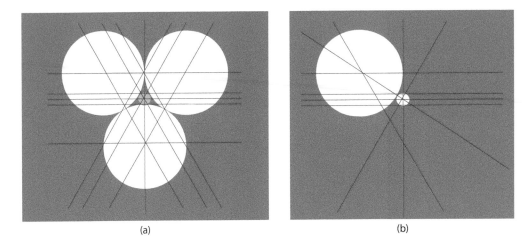

(a)   (b)

FIGURE 7.4 *(a) Three bubbles in a hexagonal packing. (b) How to calculate the size of a small bubble between three larger ones.*

This is because the triangle whose vertices are the centers of large bubbles make an equilateral triangle, so that the angle is $180/3$, *i.e.*, $60°$. As the line from the center of one bubble to the center of the small center is dividing this angle by two, we see a rectangular triangle and the radius of the small bubble can be calculated. And the new density could be determined. I leave that to you.

But let us finish by a more gourmand addition. When you make a pie dough, you meet the same question but here you have starch granules instead of air bubbles, and butter instead of the liquid of the foam. The calculation made earlier about cubes can tell you that you need one part of butter for one part of flour, which is the real recipe.

7.3 HOW THICK IS THE LIQUID LAYER BETWEEN TWO OIL DROPLETS IN A MAYONNAISE OR BETWEEN TWO AIR BUBBLES IN A WHIPPED EGG WHITE?

7.3.1 An Experiment to Understand the Question

7.3.1.1 First, the General Idea

Microscopy is so simple and so useful that one microscope should be on each kitchen bench: there are so many wonderful landscapes to observe! No need to go to Mars or elsewhere, and, as the American physicist Richard Feynman said, "there is plenty of room at the bottom".

Here, we want to observe first the objects that we are dealing with.

1. In a large vessel, place an egg white, and use a whisk to make a foam;

2. use the tip of a knife or of a spatula to take a tiny sample of the foam (*e.g.*, some cubic millimeters) that you deposit on a glass slide;

3. put the glass slide on the stage, and secure it with the stage clips;

4. turn the light on, under the slide;

5. select the objective lens giving the lowest magnification;

6. move the focus knob until the image comes into focus;

7. take the vessel with the foam, and add a spoon of oil;

8. whip;

9. add more and more oil while whipping, and take samples at regular intervals; observe them using the microscope.

7.3.1.2   Questions of Methods

We already discussed the microscopic picture of whipped egg (Figure 7.5), but now we can observe how it changes with the introduction of oil (Figure 7.6).

When this experiment is performed, the foam is transformed first in an emulsified foam, and then in an aerated emulsion.

7.3.1.3   Where the Beauty Lies

Here the "beauty" will be in calculation: when one looks to the two previous pictures, it is clearly impossible to get an idea of the size of the liquid membrane between air

FIGURE 7.5   *Here, we see only air bubbles: they are easy to recognize because of their "thick" wall, which is due to optical effects.*

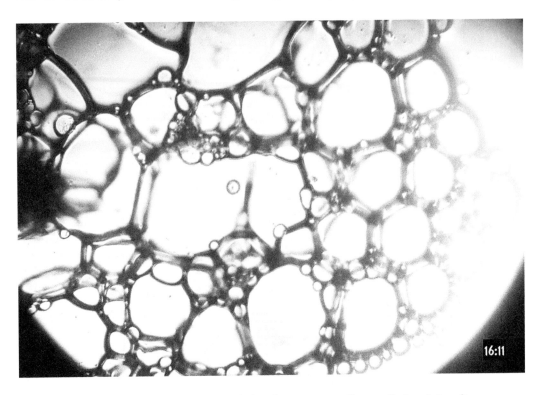

FIGURE 7.6 *When air bubbles and oil droplets are together, oil droplets often appear smaller, but the main difference is their thinner wall.*

bubbles, for foams, or of the size of the aqueous solution between oil droplets when emulsions are concerned: optical effects prevent this determination. On the contrary, we shall see now how easy it is to get an idea of such a distance using a simple calculation.

7.3.2 The Question

This exercise is designed to stress the similarity of foams and emulsions, but it also shows how to simplify the systems that one tries to characterize quantitatively.

7.3.3 Analysis of the Question

7.3.3.1 The Data is Introduced

For this question, no numerical data is provided, but you need to know how to make a mayonnaise and how to whip an egg white.

Mayonnaise sauce can be obtained traditionally from one egg yolk (about half an egg, *i.e.*, 30 g), with one spoon of vinegar (let us say 15 g).

Egg yolks (about 30 g) are made of 50% water (15 g), 35% lipids (2/3 triglycerides and 1/3 phospholipids, 10 g as a whole) and 15% proteins (5 g) (Anton, 1998).

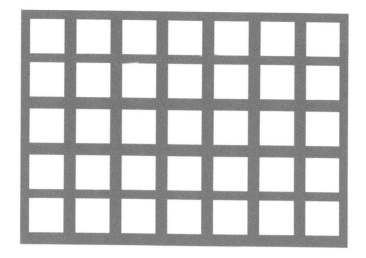

FIGURE 7.7 *A model (lattice) for a foam or for an emulsion.*

Oil is added while whisking until a thick emulsion is formed (about 300 g of oil for one egg yolk).

For whipped egg white, we know now well that the mass is about 30 g, with 90% water (27 g), and 10% proteins (3 g).

7.3.3.2 Qualitative Model

For both systems, spherical objects (gas bubbles or oil droplets) of radius r (in the assumption of a monodisperse, *i.e.*, only one size, system) are considered.

For simplicity, the dispersed objects are considered as cubes in a square lattice, with water in the middle (Figure 7.7).

For the emulsions, proteins and phospholipids can be found either dissolved, or as micelles in the aqueous phase, or at the water-oil interface.

For the foam, the proteins are at the air-water interface, or in solution.

7.3.3.3 Quantitative Model

We make the calculation for an emulsion. Let it be:

- V the total volume.

- V_o the volume of oil.

- c the side of elementary cubes of oil.

- V_w the volume of water.

- ε the thickness of the liquid phase between two dispersed structure.

7.3.4 Solving

7.3.4.1 *Looking for a Solving Strategy*

It was observed many times that the students found it difficult to identify a calculation strategy. In order to make it easier, it is useful to remember that in this part of the calculation, we almost do the calculation, but only "in principle"; *i.e.*, with words, avoiding diving into algebra. The "way" should be very clear, and why not make a numbered list, so that you can be even more confident? Let us try:

1. We start by considering that the system is made of oil cubes organized as a cubic grid;

2. water is evenly distributed between the oil cubes;

3. we know the amount of water and the amount of oil but we miss the number of oil cubes;

4. because we assumed the size of oil cubes, we can calculate the mass (or volume) of an individual oil cube;

5. and with the total mass of oil, and the mass of one oil cube, we can determine the number of oil cubes;

6. the amount of water on each cube will be the quantity of water divided by the number of cubes;

7. and from this quantity, we shall determine the thickness around each cube;

8. because this amount makes a layer the volume of which is equal to the area of the oil cubes multiplied by the thickness of the layer, we now need the area of the oil cubes to determine the thickness; but this is easy because we assumed the size of the cubes;

9. and finally, because the cubes are one against the other, the real thickness that we are looking for is double the thickness of water on each cube.

In the end, we shall check for the thickness that should be due to phospholipids and proteins.

7.3.4.2 *Implementing the Strategy*

The strategy is so clear that we shall now only translate the sentences into equations. But let us repeat our strategy and insert the equations:

1. We start by considering that the system is made of oil cubes organized as a cubic grid : nothing to add, algebraically.

2. Water is evenly distributed around the oil cubes: here also, nothing to do.

3. We know the amount of water and the amount of oil, but we miss the number of oil cubes: V_w is the volume of water, and V_o is the volume of oil. By the way, we do not know initially the volumes, but only the masses, so that we need here the densities to make the correspondence:

$$\varrho_o = \frac{M_o}{V_o} \xrightarrow{\text{isolate for V[o]}} V_o = \frac{M_o}{\varrho_o}$$

$$\varrho_w = \frac{M_w}{V_w} \xrightarrow{\text{isolate for V[w]}} V_w = \frac{M_w}{\varrho_w}$$

Here, we used the software to find the required values, but, as said earlier, this is only a way to validate our own determination: two precautions are better than one.

4. Because we assumed the size (side c) of oil cubes, we then can determine the volume v of an individual oil cube:

$$v = c^3.$$

5. With the total volume of oil, and the volume of one oil cube, we can determine the number N of oil cubes:

$$V_o = N \cdot v = N \cdot c^3$$

$$N = \frac{V_o}{c^3}$$

6. The volume of water attributed to each oil cube will be the quantity of water divided by the number of oil cubes:

$$v_w = \frac{V_w}{N} = \frac{V_w}{\left(\frac{V_o}{c^3}\right)} = \frac{V_w c^3}{V_o}$$

7. And from this quantity we shall determine the thickness around each cube: here, there is nothing to do because this is simply the announcement for the next steps.

8. Because this volume makes a layer the volume of which is equal to the area of the oil cubes multiplied by the thickness of the layer, we now need the area of the oil cubes to determine the thickness. But this is easy because we assumed the side of the cubes. Let a be the area of an oil cube; it is six times the area of a face, or:

$$a = 6 \cdot c^2$$

If ε is the thickness of water on one cube, the volume of water distributed at the surface on each cube is equal to:

$$v_w = a \cdot \varepsilon$$

But remember the former expression. Using it here, we have the equation:

$$v_w = \frac{V_w c^3}{V_o} = a \cdot \varepsilon = 6c^2 \cdot \varepsilon$$

From it, we determine ε:

$$\varepsilon = \frac{V_w \cdot c}{6 \cdot V_o}$$

9. And finally, because the cubes are one against the other, the real thickness e that we are looking for is double the thickness of water on each cube:

$$e = 2 \cdot \frac{V_w \cdot c}{6 \cdot V_o} \xrightarrow{\text{simplify symbolic}} e = \frac{V_w c}{3 V_o}$$

But remember that the initial data is the masses of oil and water, not the volume. We have to use the expressions written in step #3:

$$e = \frac{V_w c}{3 \cdot V_o} = \frac{c \cdot \left(\frac{M_w}{\varrho_w}\right)}{3 \cdot \left(\frac{M_o}{\varrho_o}\right)} = \frac{c M_w \varrho_o}{3 \varrho_w M_o}$$

7.3.4.3 Discussion, Validation

First, let us observe that the expression that we found has the dimension of c; *i.e.*, a distance, as it has to have.

We also observe that the thickness of the liquid lamellae increases with the quantity of water, and this should be indeed the case. It decreases with an increasing volume of oil, and certainly if one increases the oil, the water has to distribute in a bigger quantity.

Densities are present in the results because we considered masses instead of volumes.

7.3.5 Expressing the Results

7.3.5.1 The Formal Result Found Is Written Down Again

As usual, this sub-section is always the easiest:

$$e = \frac{c M_w \varrho_o}{3 \varrho_w M_o}$$

7.3.5.2 Finding Digital Data

We have:

- the mass of water is from egg yolk (15 g) and vinegar (15 g);
- the mass of oil is from oil (300 g) and lipids from the yolk (10 g);

- the densities can be decided as 1000 kg/m³ for water, and 800 kg/m³ for oil (for oil, we gave previously more precise values, but adding digits will not change the order of magnitude of the result);

- the size of structures is known from microscopic studies, and is between 0.001 and 1 mm; we start with a value of 0.1 mm.

7.3.5.3 Introduction of Data in the Formal Solution

As always, we use the International System of Units:

$$subs\left(M_w = 30 \cdot 10^{-3}, c = 10^{-4}, \varrho_o = 800, \varrho_w = 1000, M_o = 0.3, \frac{cM_w\varrho_o}{3\,\varrho_w M_o}\right)$$

$$2.67 \times 10^{-6}$$

This value is in m: we find it to be of the order of the micrometer.

7.3.5.4 Discussion, Playing with Parameters in Order to Explore the Solution Space

First, we have to observe that the value that we found is of the order of magnitude of that which can be observed under the microscope.

Then, we have to find the extreme values:

$$subs\left(M_w = 30 \cdot 10^{-3}, c = 10^{-6}, \varrho_o = 800, \varrho_w = 1000, M_o = 0.3, e = \frac{cM_w\varrho_o}{3\,\varrho_w M_o}\right)$$

$$e = 2.67 \times 10^{-8}$$

$$subs\left(M_w = 30 \cdot 10^{-3}, c = 10^{-3}, \varrho_o = 800, \varrho_w = 1000, M_o = 0.3, e = \frac{cM_w\varrho_o}{3\,\varrho_w M_o}\right)$$

$$e = 2.67 \times 10^{-5}$$

And what about making a curve of the thickness as a function of the side of oil cubes, instead of calculating values one after the other? I leave this to you.

7.3.5.5 Validation

For this chapter, let us try to make the calculation entirely new, using now the example of a foam, instead of an emulsion.

Let V be a volume of whipped egg white (Figure 7.8).

There are N bubbles of side c, so that the volume of each bubble is $v = c^3$.

The volume V is equal to the volume of bubbles plus the volume of the liquid v_l.

$$V = N \cdot c^3 + v_l$$

As we know V, we also know c^3, and v_l through the density of egg white:

$$\varrho = \frac{m}{v_l}$$

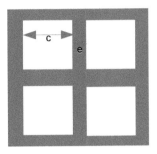

FIGURE 7.8 *A whipped egg white as modeled as a grid of cubic air bubbles disperses in an aqueous solution (blue).*

Or:

$$v_l = \frac{m}{\varrho}$$

$$solve(V = N \cdot c^3 + v_l, N)$$

$$\frac{V - v_l}{c^3}$$

Or:

$$N = \frac{V - v_l}{c^3} = \frac{V - \left(\frac{m}{\varrho}\right)}{c^3}$$

If the volume of liquid is distributed around the N bubbles, this means that it makes a volume ν around each bubble:

$$\nu = \frac{v_l}{N} = \frac{\left(\frac{m}{\varrho}\right)}{\frac{V - \left(\frac{m}{\varrho}\right)}{c^3}}$$

This volume makes a layer equal to six smaller layers (one per side), *i.e.*:

$$\nu = 6 \cdot c^2 \cdot \left(\frac{e}{2}\right)$$

We can now determine e:

$$solve\left(\nu = 6c^2 \cdot \left(\frac{e}{2}\right), e\right)$$

$$\frac{\nu}{3c^2}$$

Finally, we get:

$$e = \frac{\nu}{3c^2} = \frac{\left(\dfrac{\left(\frac{m}{\varrho}\right)}{V-\left(\frac{m}{\varrho}\right)}\right)}{c^3}\Big/ 3c^2$$

And we immediately make a numerical application:

$$subs\left(\varrho = 1000, m = 30e-3, V = 1e-3, c = 10^{-5}, e = \frac{\left(\dfrac{\left(\frac{m}{\varrho}\right)}{V-\left(\frac{m}{\varrho}\right)}\right)}{c^3}\Big/ 3c^2\right)$$

$$e = 1.030927835\ 10^{-7}$$

Here, we get the same kind of result as before, but indeed this is the same calculation formally.

7.3.6 Conclusions and Perspectives

There are many ways to go further with this calculation. One way could be to consider one assumption that we made, about neglecting the thickness of the interfaces, with lipids and proteins. For example, we can start from the expression of the volume of water around each cube, and distinguish the various contributions of oil and lipids. The same kind of question could also be asked for proteins.

7.4 WHAT IS THE MAXIMUM VOLUME OF WHIPPED EGG WHITE THAT ONE CAN MAKE FROM ONE EGG?

7.4.1 An Experiment to Understand the Question

7.4.1.1 First, the General Idea

Let us begin by exploring the whipping of an egg.

1. One week in advance, we weigh an egg white (about 30 g), and we put it in a vessel, at room temperature, or in an oven at a temperature less than 60°C (if you do this, you can also place a whole egg, in its shell, for comparison). This will show that the 30 g of the egg white can be reduced to about 3 g: the residue is a glassy yellow material, and it is almost entirely made of proteins. If you did the experiment with the whole egg in an oven, you would be able to observe that a noise is heard when you shake the dried egg, and if you open it, you will see that the yolk has shrunk by 50%, and that the egg white forms a thin (remember: 10% proteins only) layer of "resin" on the outside of the yolk.

2. Having observed that an egg white is made of water (that evaporated during drying) and proteins, then take a clean vessel, such as a large salad bowl, add pure water, and whip using a clean whisk: you observe that air bubbles are introduced, and the appearance become somehow white.

3. When you stop whipping, you see the air bubbles moving quickly upwards, and after some seconds they explode at the surface, while the liquid become colorless and transparent again.

4. Now, compare that with the whipping of an ordinary egg: you put an egg white in the vessel, and you begin observing that it is transparent, but with a yellow/green color (so it does not deserve the name "egg white").

5. Rapidly push a pen or a metal wire into the egg white: you observe now that some air bubbles are introduced. They are quite stable: can you determine how much time they remain? By the way, as the difference between water and egg white is simply the presence of proteins, we have to conclude that the stabilization of air bubbles in whipped egg white is due to proteins.

6. If you use now two pens or two wires to push air, you observe that you get about twice the number of air bubbles (you can count them).

7. Now, take a whisk and count the number of wires: the more you have, the faster the egg white will be whipped because more air bubbles will be introduced into the egg white.

8. After some minutes, you get white firm foams that you divide into two bowls.

9. For one half, go on whipping rapidly, even after the firm foam is made: after some minutes, it is likely that you will be able to observe small white dots, the nature of which is still unknown (coagulation of proteins?). During this whipping, you can try to observe that the volume does not increase much after a maximum volume is reached.

10. Before using the second half, let us ask ourselves the question: we had an egg white, made of water and proteins, and we whipped, with a volume expansion; after some time, the expansion stopped, but why? In the egg white, we have water, proteins, and air, so that we can conclude that "something is missing", but what? It can be either water, or proteins, or air (or two at the same time, but for probabilistic reasons, this is unlikely). Indeed, air is not missing: there is plenty of it around... and whipping two egg whites generates twice the volume of foam. Then are we missing water or proteins?

Because the simplest test to make is to add water, let us use the second half of the previous whipped egg white, and let us add a spoon of water: whipping creates more volume. Then, let us add more water, whip, add more water, whip again... and you will be interested to learn that doing this with children during an event in Paris, we got more than 40 L of whipped egg white after two and a half hours of whipping (27th September 2012).

7.4.1.2 Questions of Methods

Here, there are four main comments to make, about the previous experiment.

1. First, about the color of egg white, we said that it was transparent, yellow and greenish. But these are three adjectives, and one has to remember that it is "forbidden" to use adjectives (and adverbs) in science: we have instead to answer the question "how much?".

   For transparency, it is not difficult to use a mobile phone and to record the intensity of the light that you get in the two different conditions:

   - through a glass, with a light being shone in the line light-empty glass-mobile
   - through a glass with egg white, with the same arrangement.

   For the color, again the mobile phone can be used for taking pictures that are analyzed with software such as *ImageJ* or *Maple*, for example.

2. Now we have to discuss the technological question of the number of wires in a whisk. It is good to know that, as said above, in the past the whisks were made out of wicker, with simple lashes grouped together on one hand. However, when you raised up such a tool, you got projections in the face, and someone had the idea to fold the lashes and transform the tool into the modern whisk. Later, wicker was replaced by metal, recently by the wonderful stainless steel.

   Now, if you remember the previous analysis, you understand that the efficiency depends on the number of wires: the more numerous they are, the more air bubbles are pushed in the liquid. And of course, using two whisks together is twice as fast as than using one.

   This leads to a small funny calculation: imagine that a wire of a whisk is 1 mm wide, and 10 cm long, and such a wire pushes air bubbles of 1 mm^3 in diameter. The number of such bubbles would be 100 for one wire, and about 10^3 for a whisk with 10 wires. If you whip for 100 seconds (more likely than 10 s), with a velocity of one circle per second, this makes an order of magnitude of 10^5 bubbles of 1 mm^3 each, *i.e.*, 10^5 mm^3, or 0.1 dm^3 (0.1 L), which is the right order of magnitude for the volume of an ordinary whipped white.

3. About whipping egg whites, it is often said that it is useful (some say that the goal is firmness, others say that it is for stability, and others do not explain why) to add a pinch of salt, or some lemon juice, but how is it really useful?

   Let us focus on the first question about salt: you can try to explore this experimentally, but mind that measuring the volume of foams is very difficult, in particular because they can trap large pockets of air in the vessels where they stay or where they are decanted for measurement. And in the end, you would have to compare the quantity of foam with salt and the quantity of foam without salt. Because such quantities are known with large uncertainties, it would be very difficult to arrive at a conclusion.

   And here, it is worth reminding students that results such as the one in Figure 7.9 have no value.

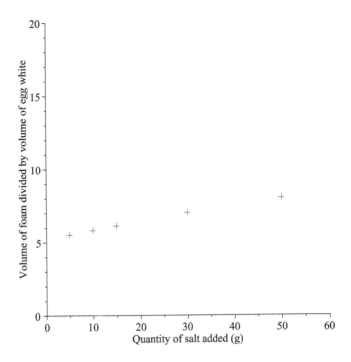

FIGURE 7.9 *In this scheme the axes are labeled and the units are given, but the graph has no value because the uncertainties are not given. In particular, one does not know anything about the possible meaning of the size of symbols.*

Imagine now that the uncertainties were determined, with the following improved representation showing uncertainties (Figure 7.10).

Now, this is interesting because, instead of the red line, we could also have a decreasing line (Figure 7.11).

4. The last methodological comment about our experiments of whipping eggs will be metaphorically explained (with a smile), taking the example of "teaching", and demonstrating why teaching is "impossible". Imagine that a teacher presents a course with a velocity V, and that there are different students in the class. Let us imagine that one student can understand at velocity v. Then, the probability that the student can follow is nil.

Indeed, a probability is defined as the number of occurrences (here $1 : V = v$) over the number of possibilities (infinite : v being chosen as fixed, V can take any real value in a continuum). And indeed, most frequently, students are lost, or they are bored.

7.4.1.3 *Where the Beauty Lies*

After the previous experiments, we know that we can make very large volumes of foams, from only one egg. But in the experiment it was proposed to add water, and

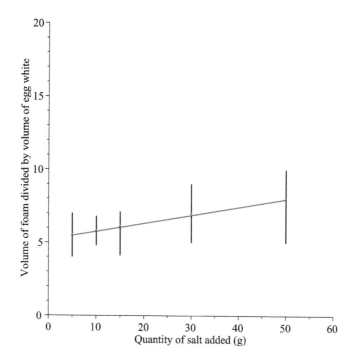

FIGURE 7.10 *An improved representation of the same data, showing the measurement uncertainties. The red line is from the first to the last measurement.*

the culinary virtue of water is poor. In kitchens, chefs generally use instead aqueous solutions that provide more flavor such as stocks, broths, juices, wine, etc.

Indeed, the recipe of culinary systems called "geoffroys" (from the name of a famous French dynasty of pharmacists/chemists of the 17–19th centuries) is to whip an egg white until it is firm. Then, you alternate the addition of apple juice and sugar, while whipping, so that you get the needed volume of foam: the geoffroy that you get has the flavor of the juice.

To turn this geoffroy into a "vauquelin" (another French chemist, of the 18th century), you simply cook the geoffroy in a microwave oven until you see an expansion by about 30%: this demonstrates that water evaporation (or rather boiling) occurred, and accordingly that a temperature of 100°C was reached, at which you are sure that the proteins coagulated. The system is now a solid foam.

7.4.2 The Question

Let us now examine the initial question given in the title of the chapter. Certainly, experimental work could produce a lot of foam, but how much? Indeed, the experience with children, for beating the world record of whipped egg whites, was based on the calculation that you are invited to make here.

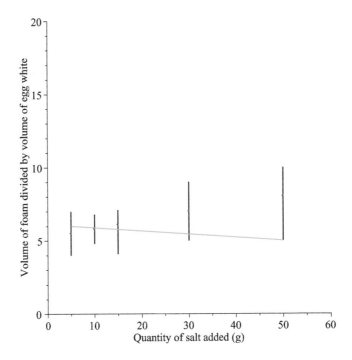

FIGURE 7.11 *With the same data used for Figure 7.10, a decreasing trend can be drawn: this figure shows that the particular experimental setup used for this experiment should be modified. Also, one should remember that the comparison of two digital results has to be analyzed statistically, using methods such as Student's or ANOVA (analysis of variance).*

7.4.3 Analysis of the Question

7.4.3.1 The Data is Introduced

Here one simply needs to know that an egg white has a mass of about 30 g, comprising 10% proteins and 90% water.

7.4.3.2 Qualitative Model

Let us begin with a scheme (Figure 7.12).

Cooks know that the volume of whipped egg white increases during whipping, until a maximum volume is reached. At that point, we could ask whether the expansion is limited by water or by proteins. Indeed, the experiment shows that water is needed, and this is obvious from microscopic experiments (air bubbles are closely packed), so that it is worth calculating how much of foam could be obtained using the protein content of an egg white.

We shall assume that all air bubbles are spherical, and of the same diameter.

FIGURE 7.12 *Envisioning a foam. One should try to make this as real as possible, but of course the scale should be chosen first. As a rule, the main description should come before a more detailed one. This allows one to make assumptions.*

7.4.3.3 Quantitative Model: Describe Formally the Characteristics of the Objects That You See on Your Picture, i.e., Introduce Formal Symbols That Will Be Used

Let it be as follows:
   V the volume of foam,
   r the radius of air bubbles,
   m the initial mass of egg white,
   v the volume of an egg white.

7.4.4   Solving

7.4.4.1   Looking for a Solving Strategy

Firstly, we shall calculate how much volume of foam can be obtained by simply whipping, assuming close packing of air bubbles all of the same size (1).

Secondly, we shall calculate how many such bubbles one can obtain if the proteins form a monolayer at the air-water interface (2).

Thirdly, we shall consider the case of only one bubble forming a "one-bubble whipped egg white", and we shall calculate the radius of such "one bubble foam" (3).

7.4.4.2   Implementing the Strategy

1. Close packing of spherical air bubbles all of the same size

First, let us imagine that air bubbles are simply closely packed (we have to keep this assumption in mind, for discussing the results later, because microscopic pictures show that this is not true). In this assumption, let us assume that the air bubbles occupy of proportion p of the total volume V of the foam (it is the "density"); the liquid occupies $(1 - p) \cdot V$.

As we know the volume v (given by $\varrho = \frac{m}{v}$) of an egg white, we can calculate:

$$\frac{m}{\varrho} = (1 - p) \cdot V$$

So that the volume V is:

$$V = \frac{m}{\varrho \cdot (1 - p)}$$

2. How many such bubbles can we obtain if the proteins form a monolayer at the air-water interface

We have now to make a new assumption: in an extreme case, we consider one layer of protein at the air-water interface. Using this assumption, we start from the protein content: a mass m. This corresponds to a number n of moles:

$$n = \frac{m}{MM}$$

where MM is the average molar mass for egg white proteins.

And the number N of proteins is:

$$N = \frac{m}{MM} \cdot N_A$$

where N_A is the Avogadro number.

For each protein, there is an area a. So that the total area covered by proteins is:

$$A = Na = \frac{m}{MM} \cdot N_A \cdot a$$

Now, we assume that air bubbles have a radius r. The area of one bubble is:

$$a_b = 4\pi r^2$$

The number N_b of bubbles can be determined by dividing the total area by the area of one bubble:

$$N_b = \frac{A}{a_b} = \frac{\frac{m}{MM} \cdot N_A \cdot a}{4\,\pi\,r^2} = \frac{m \cdot N_A \cdot a}{MM\,4\,\pi\,r^2}$$

From which we can determine the total volume of bubbles, as the product of the number of bubbles by the volume of one bubble:

$$V_b = N_b \cdot v_b = \frac{m \cdot N_A \cdot a}{MM\,4\,\pi\,r^2} \cdot \frac{4}{3} \cdot \pi\,r^3 = \frac{m \cdot N_A \cdot a \cdot r}{3\,MM}$$

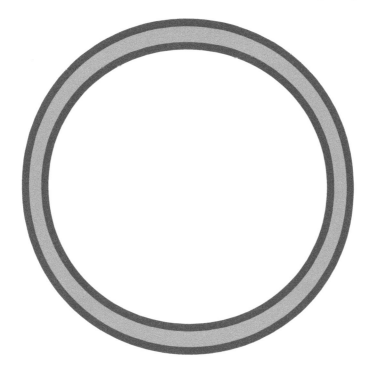

FIGURE 7.13 *A model for an only-one-bubble whipped egg white. The liquid phase (blue) is between two layers of proteins (red).*

3. Only one bubble forming a one-bubble whipped egg white.

Here the calculation is almost done, because we calculated in (2) the total area covered by all proteins of an egg white. Now, we shall have however a new model (Figure 7.13).

Here, we assume a protein layer at the external air-water interface, and another one at the inner water-air interface.

If R is the radius of the large bubble, its area is $4\pi R^2$. With our assumptions, and assuming that the layer of water is thin compared to the radius (this means writing $R - e \approx R$, with e the thickness of the water layer), the proteins have to distribute on the double of this area, *i.e.*, $8\pi R^2$.

Remember for the previous paragraph that the total area covered by proteins is:

$$A = \frac{m}{MM} \cdot N_A \cdot a$$

where a is the area covered by one protein.

Writing that the total area covered makes the double layer of proteins

$$8\pi R^2 = \frac{m}{MM} \cdot N_A \cdot a$$

We can deduce the radius of the large bubble:

$$R = \sqrt{\frac{m N_A a}{8\pi MM}}$$

But of course, we can make such a big bubble only if there is enough water.

7.4.5 Expressing the Results

7.4.5.1 Finding Digital Data

An egg white has a mass of 30 g, with 90% water, *i.e.*, 27 g of water, and 3 g of proteins.

Ordinarily, air bubbles have a radius between 0.01 and 1 mm.

For (2), we also need the area of one protein, and we have seen (Section 7.2) that the diameter of a fully coiled protein is 1.5×10^{-9} m, whereas the length of a fully extended protein is 2×10^{-8} m.

In the first case, the area would be the square of this value, and in the second case we can multiply the length of the extended protein by a width that can be assumed to be 5 covalent bonds long ($2 \times 10^{-8} \cdot 5 \cdot 1.5 \times 10^{-10} = 1.5 \times 10^{-17}$). We shall choose the longest value first.

7.4.5.2 Introduction of Data in the Formal Solution

1. With the close packing assumption:

$$subs\left(\varrho = 1e3, m = 30e - 3, p = 0.74, V = \frac{m}{\varrho \cdot (1 - p)}\right)$$

$$V = 1.15 \times 10^{-4}$$

Remember that because we used the units of the International System, the result is in m^3; *i.e.*, the volume is about one tenth of a liter.

2. With proteins limiting the volume, we have two cases.

First, when the proteins are considered fully coiled, then $a = (1.5e - 9)^2$

$$subs\left(m = 3e - 3, a = (1.5e - 9)^2,\right.$$

$$\left. N_A = 6e23, r = 1e - 4, MM = 45, p = 0.74, \frac{m \cdot N_A \cdot a \cdot r}{3\,MM}\right)$$

$$3.00 \times 10^{-3}$$

This means about 3 L.

Now, considering fully extended proteins, $a = 1.5 \times 10 - 17$, and

$$subs\left(m = 3e - 3, a = 1.5e - 17,\right.$$

$$\left. N_A = 6e23, r = 1e - 4, MM = 45, p = 0.74, \frac{m \cdot N_A \cdot a \cdot r}{3MM}\right)$$

$$2.00 \times 10^{-2}$$

This is of course much bigger (about 20 L).

3. For one large bubble, assuming fully coiled proteins, the radius would be:

$$evalf\left(subs\left(m = 3e-3, a = (1.5e-9)^2, N_A = 6e23, MM = 45, R = \sqrt{\frac{m\,N_A\,a}{8\,\pi\,MM}}\right)\right)$$
$$R = 1.8923493910$$

A radius of about 2 m!

And in the second assumption:

$$evalf\left(subs\left(m = 3e-3, a = 1.5e-17, N_A = 6e23, MM = 45, R = \sqrt{\frac{m N_A a}{8\pi MM}}\right)\right)$$
$$R = 4.8860251170$$

This (about 5 m) would be more.

Now, would you like to calculate the amount of water needed, in these various cases, assuming a monomolecular layer between the proteins?

7.4.6 Conclusions and Perspectives

The last calculation seems silly, and it is true that big bubbles are fragile, but on the other hand, large bubbles are made when whipped egg whites are produced by a pump blowing in the liquid.

And now that the idea is given, we could open a contest: what will be the biggest bubble that you can make?

REFERENCES

Anton M. 1998. *Structure and functional properties of hen egg yolk constituents,* Recent Research Developments in Agricultural and Food Chemistry. 2 <hal-02838267>.

De Gennes PG. 1998. *L'intelligence en physique,* Pour la Science, 254, 14.

FAO. 2016. *Pulses,* https://www.fao.org/pulses-2016/news/news-detail/en/c/337107/, last access 2022-01-10.

The year 2016 was chosen to be the FAO international year of pulses because these plants can be very important for sustainable human food: they contain proteins (and you remember that human beings are made of about 70% water and 20% proteins). Certainly, meat is overconsumed in some countries, but proteins are needed nonetheless.

Hales T, Adams M, Bauer G, Dang TD, Harrison J, Hoang LT, Kaliszyk C, Magron V, McLaughlin S, Nguyen TT, Nguyen QT, Nipkow T, Obua S, Pleso J, Rute J, Solovyev, A, Ta THA, Tran NT, Trieu TD, Urban J, Vu K, Zumkeller R. 2017. *A Formal Proof of the Kepler Conjecture,* Forum of Mathematics, Pi. 5: e2. doi:10.1017/fmp.2017.1.

IUPAC. 2019. *Foams*, Compendium of Chemical Terminology, https://doi.org/ 10.1351/goldbook, last access 2022-01-10.

IUPAC, IUPAC, again IUPAC: this wonderful entreprise can avoid divisions between humans; in a way, it is a way to fight the "Babel tower" syndrome.

Langevin D. 2016. *Aqueous foams and foam films stabilised by surfactants*, Comptes rendus mécanique, 345(1), 47–55.

Le Floch-Fouéré C, Beaufils S, Lechevalier V, Nau F, Pézolet M, Renault A, Pezennec S. 2010. *Sequential adsorption of egg-white proteins at the air–water interface suggests a stratified organization of the interfacial film*, Food Hydrocolloids 24, 275–284.

Sun Y, Jin H, Sun HH, Sheng L. 2020. *A Comprehensive Identification of Chicken Egg White Phosphoproteomics Based on Novel Digestion Approach*, Food Chemistry, DOI: 10.1021/acs.jafc.0c03174.

You see, I am trying to give you the most recent information. But the sources can evolve in the future.

This H. 2021. *Masterclass*, https://www.youtube.com/watch?v=XX8P9z5GSlY, last access 2022-01-10.

There are six masterclasses online about molecular and physical gastronomy. More to come.

This vo Kientza H. 2021. *Disperse system formalism*, Handbook of Molecular Gastronomy, CRC Press, 207–212.

CHAPTER 8

Emulsions (L1/L2)

N ow, you should be able to interpret the L1/L2 from the title of the chapter, especially if you have the IUPAC definition of "emulsions":

A fluid colloidal system in which liquid droplets and/or liquid crystals are dispersed in a liquid. The droplets often exceed the usual limits for colloids in size. An emulsion is denoted by the symbol O/W if the continuous phase is an aqueous solution and by W/O if the continuous phase is an organic liquid (an 'oil'). More complicated emulsions such as O/W/O (*i.e.* oil droplets contained within aqueous droplets dispersed in a continuous oil phase) are also possible. Photographic emulsions, although colloidal systems, are not emulsions in the sense of this nomenclature.

(IUPAC, 2019a)

Here you see that our L1/L2 is broader than O/W (oil dispersed in water), or W/O, because there are other possibilities than the three given in the IUPAC definition. In our L1/L2 expression, L1 stands for any liquid phase, and L2 as well.

8.1 WHY ONE HAS TO WHIP OIL TO MAKE A SAUCE MAYONNAISE?

8.1.1 An Experiment to Understand the Question

8.1.1.1 *First, the General Idea*

Sauces are fascinating systems because they are often of colloidal nature, a category of physical and chemical systems that comprises emulsions, such as mayonnaise. About mayonnaise precisely, it appeared only in the 19th century and evolved from a previous sauce called "remoulade", which was already discussed in the 1319 book entitled *Le Viandier*, by the French Guillaume Tirel (nickname Taillevent), who was the cook for several kings of France (the book based on a previous version, dated 1300). Remoulade was based on mustard and various liquids, hot or cold, including oil (Taillevent, 2000).

Later, cooks added egg yolks to remoulade because it has been well known in the kitchen (and obvious if tasted) that the flavor of the egg yolk is really good.

And suddenly, after 1800, we observe the appearance in culinary books of the sauce mayonnaise, in which there is no mustard, and instead the emulsion is obtained thanks to the yolk (Carême, 1828).

The sauce was initially called "magnonnaise", or "mahonnaise", and it is said that the first name was because one has to "manier" (to mix, in French) and the second name was given because the sauce was discovered by the cook of the cardinal of Richelieu, during a war event in Port Mahon (the capital city of Majorca, in Spain). Indeed, nobody knows the exact origin of the sauce, but it is a fact that an emulsion similar to aioli (you grind garlic with olive oil) was already present in Marin's *Les dons de Comus*, published in 1742 under the name "beurre de Provence" (butter from Provence) (Marin, 1758).

In order to make a mayonnaise (no mustard, remember):

1. Put an egg yolk in a vessel.

2. Add a spoon of vinegar or another liquid (*e.g.*, stock, lemon juice, etc.).

3. Whip, while adding oil slowly until a thick preparation is obtained.

4. Add salt and pepper to taste.

8.1.1.2 *Questions of Methods*

Mayonnaise can "fail", that is the oil and aqueous phase can separate, with oil floating over an aqueous medium. About this, many interpretations were proposed: the sauce would fail at cold temperatures; the sauce would fail at hot temperatures; the sauce would fail if women in the room had their periods; the sauce would fail if the moon was full; the sauce would fail if there was some wind; the sauce would fail if whipped insufficiently; the sauce would fail in front of your mother-in-law; the sauce would fail if... (This, 2010).

I tested and refuted many of these assumptions. And finally, only physical chemistry provides the key for the success: if, at the beginning of the whipping process, you have more aqueous phase than oil, the whisk will divide the oil into droplets that will be dispersed in the aqueous phase, and the sauce will be successful as long as you go on dividing the oil in droplets that are dispersed in the emulsion; on the contrary, if you add too much oil, the whip will disperse the aqueous phase in the oil, and this will make an very unstable system because the "surfactant" (proteins or phospholipids) curve the oil-water interface in the wrong direction (in other words, the interface is preferably bent so that there are oil droplets in water) (De Gennes *et al.*, 2004).

8.1.1.3 *Where the Beauty Lies*

Up to now, I have not discussed much of technology, except for examples such as the number of wires making up a whisk. But let us remember (always, always, always) that human activities are defined by a goal and the way leading to it. Indeed a good piece of advice, before beginning any activity is to ask for a clear goal: what is all this about? what is the goal? what is the use of it? where do we want to go? what do we want to do and why? Only when the goal is clear, can we try to reach it.

This is very important for culinary activity because most of the time, processes and recipes have not been based on this kind of analysis, but rather on reproduction of knowledge that was found empirically, and not interpreted in terms of chemistry or physics. In particular, about emulsions, one has to know that where cooks use whisks, physical chemists use ultrasonic devices, such as probes (a metal rod) vibrating with high energy and frequency: when dipped in a mixture of oil, water and surfactant, this efficiently divides the oil and disperses it into water, making in some seconds what whisks need minutes to produce.

I proposed to cooks the use of such ultrasonic probes as early as 1994, but it was only in 2019 that a company began selling specific systems to cooks, as part of "molecular cooking".

8.1.2 The Question

Let us now consider the question, whatever tool we use for giving energy to the ingredients of a sauce: Why do we have to give energy to make a mayonnaise sauce? The question is simple, but the analysis will show that many different solutions are possible.

8.1.3 Analysis of the Question (Often Questions Are Solved Immediately When They Are Analyzed)

8.1.3.1 The Data is Introduced

Here, one needs to know how to make mayonnaise, *e.g.* using one egg yolk, one spoon of vinegar, and about half a liter of oil.

8.1.3.2 Qualitative Model

Let us start from the culinary system: mayonnaise sauce. As mentioned, it is an oil in water (O/W) emulsion, because the oil is whipped in a mixture of egg yolk (50% water) and vinegar (90% water) (This vo Kientza, 2021).

When mixing egg yolk and vinegar, one can observe that they "mix" easily, and this can be interpreted as water mix with water, even if the egg yolk contains 35% fat (2/3 triglycerides, 1/3 phospholipids) and proteins grouped in "granules" dispersed in a plasma (a solution of proteins). Here, we neglect the granules by simply considering that mayonnaise is obtained by dispersing oil into water (This, 2009).

One solution to the question of this chapter could be to calculate the amount of energy needed for dispersing the oil, using the surface tension: you could calculate the increase of the area of the oil-water interface (imagine that you divide a large "cube of oil" in n^3 smaller cubes: can you now calculate the area increase?), and this could be compared with the energy given to the whisk.

Another solution, used here, is to try to understand the chemical reasons for this surface tension. Previously (see Section 6.6), we considered that water molecules at the surface, when water is in contact with air, are attracted by the other molecules in the inside of the liquid, and very few by the air, so that the resultant of the forces was toward the center of the liquid.

But here we have a different system, and the question is now to understand why water and oil do not mix. Many people remain at an obvious (and tautologic) level such as "oil does not mix with water because it is hydrophobic". And when they are asked why oil is hydrophobic, many explain that molecules from oil would not make bonds with water molecules, which is not true, as oxygen atoms of the ester bond have free doublets of electrons, so that they can make hydrogen bonds, on top of all the other van der Waals bonds (between dipoles, or between dipoles and induced dipoles, or between induced dipoles together). One has to be cautious in analyzing the system well, and we shall see why we have to remember that "there are wheels within wheels".

8.1.3.3 Quantitative Model

Here, we propose a two-step calculation:

1. Estimate the enthalpy change when one hydrophobic molecule is introduced in water

2. Estimate the entropy change when a hydrophobic molecule is dipped in water.

8.1.4 Solving

8.1.4.1 Looking for a Solving Strategy

For the two calculations, we shall use a lattice calculation. For the second one, we shall use the Boltzmann formula for expression entropy S in terms of the number of microscopic configurations Ω:

$$S = k_B \ln(\Omega).$$

8.1.4.2 Implementing the Strategy

1. Change in enthalpy

Let us consider a molecule of oil (triglyceride) that we put in water (Figure 8.1).

It is worth considering the chemistry of all this, and first by drawing the chemical formula of the triglyceride molecule (Figure 8.2).

If we consider triglyceride molecules, we observe that they can make van der Waals bonds with water, but also some hydrogen bonds because of free electron doublets on oxygen atoms. In order to estimate the energy change, we need to have an idea for the energies of the various chemical bonds. In decreasing order, they are van der Waals bonds < hydrogen bonds < disulfide bridges < covalent bonds < electrostatic forces. More precisely, Table 8.1 is giving numerical data and references.

And this is well depicted in Figure 8.3 (I know that this information was provided earlier, but let us focus our mind on this because it is so useful!).

For our calculation, let us use an order of magnitude of about 5e-2 kJ·mol^{-1} for each bond. Of course, with the triglyceride molecules being large ones when entirely

FIGURE 8.1 *A triglyceride molecule going into water establishes van der Waals forces with surrounding water molecules. This is energetically favorable to dissolution.*

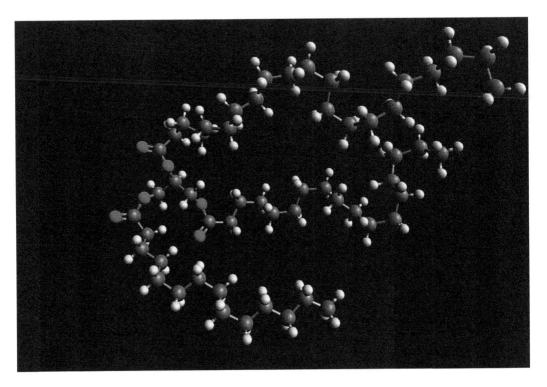

FIGURE 8.2 *Molecules of triglycerides are made of a glycerol residue, esterified by three residues of fatty acids (fatty acids are compounds that are made of a hydrocarbon chain and a carboxylic acid group); some fatty acid residues can have doubles bonds instead of single bonds between carbon atoms of their backbone (unsaturation). Mind that in this molecule, there is no glycerol* stricto sensu, *and no fatty acids as well: as in the case of the water molecules, that do not contain dihydrogen and dioxygen, atoms are missing to make real glycerol and fatty acids (chemical reactions are no physical aggregation). Hence the important use of the word "residue" that one should also use for proteins (they are not composed of amino acids, but of amino acid residues).*

unfolded, we could use the previous calculation of the distance between two water molecules to estimate the number of water molecules that could surround a triglyceride molecule extended in water.

For this, let us consider a molecule with three fatty acid residues having ten carbon atoms each. Imagine that the water molecules would wrap each residue, making a "tube" with square section, and with five water molecules long (remember that the distance between two water molecules, in liquid water, is about twice a C-C bond. This would make 20 water molecules around each fatty acid residue, and 60 for the whole triglyceride. Accordingly, the enthalpy change in dissolving a triglyceride molecule in water would be about 3 kJ · mol^{-1}.

This value is unimportant in itself, but its sign is meaningful: with our analysis, it says that (weak) bonds are created when the triglyceride molecule is introduced in water, so that it would favour the dissolution of the triglyceride molecules in water!

This is why the next calculation is so important.

Table 8.1 The Energy of the Various Forces Encountered in Physical Chemistry (Cottrell, 1958)

Force	Energy ($kJ \cdot mol^{-1}$)
Ionic	335–1050
Metallic	110–350
Covalent C-C	100–400
Disulfide bridge	75 of the energy of a covalent C-C bond
Covalent C-H	200–400
Ion-dipole	15
Hydrogen bond	8–42
Dipole-dipole	4–21
Ion-induced dipole	3
Dipole dipole	2
Induced dipole-induced dipole	0.05

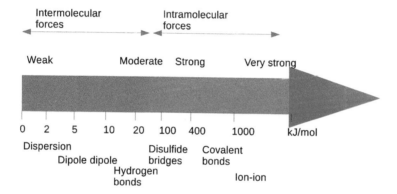

FIGURE 8.3 *The energies of the various chemical bonds.*

2. Estimation of the change in entropy

First, we have to recognize that the question of why oil and water do not mix is a complex one (Silverstein, 1998; Biedermann *et al.*, 2014): for decades, various theories have been proposed to explain a simple fact, something that should remind us of the discussion in Section 6.7 about the dissolution of sugar in water.

But here, we shall try to make a calculation using the idea that when a triglyceride molecule is introduced into water, there is an unfavorable change in entropy. Some textbooks explain that this is the result of a change in structure that takes place in the "water cage" surrounding the triglyceride molecule: water molecules would compensate for their broken hydrogen bonds by making extra hydrogens bonds with their nearest neighbors, and the "rigidity" of the cage would correspond to an entropy decrease. As:

$$\Delta G = \Delta H - T\Delta S$$

the total energy would not favor the dissolution.

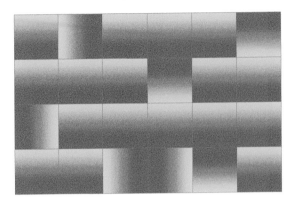

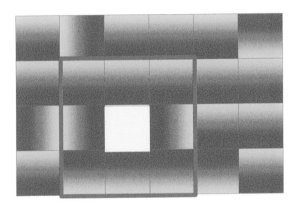

FIGURE 8.4 *Calculating the variation in entropy when a triglyceride molecule goes into water. Here, the water molecules are considered to have four possible orientations in the bulk of water. However, after a triglyceride molecule was introduced, the water molecules close to the triglyceride have restricted mobility (red square).*

However, this explanation has been criticized, and some authors have developed an alternative approach called the "scaled particle cavity theory" (Pierotti, 1976): entropy loss would stem from the decrease of rotational and translational freedom of the solvent.

Whatever the exact mechanism, remember that we want to calculate something, and this question will give us the opportunity to study the question using a lattice model again (Figure 8.4).

In the model described in Figure 8.4, we consider that the n water molecules make "cages" around the triglyceride molecules, and hydrogen bonds between the water molecules of these cages limit the mobility of water molecules: in the model that we use here, the water molecules have only one orientation, instead of four. Using this assumption, we can estimate the variation in entropy.

Before "dissolution", each of the n water molecules can be at any place, and have four possible orientations. For the first position of the "water grid", we can place any of the n water molecules with the four orientations; once this molecule is placed, the second position of the grid can be occupied by any of the $n - 1$ remaining water molecules, again with four possibilities for orientation, and so on until the entire grid is filled. The number of configurations is:

$$4n \cdot 4(n - 1) \cdot \ldots \cdot 4 \cdot 2 \cdot 4 \cdot 1 = 4n!$$

And the entropy is:

$$S_1 = k_B \cdot \ln(4\,n!)$$

where k_B is the Boltzmann constant.

After the triglyceride molecule is in water, the number of microscopic configurations is reduced. Now, let us consider that the cage is made of p water molecules. Certainly, the water molecules can occupy any place, but here, for the p positions around water, the number of orientation possibilities is reduced to 1. This means that the previous expression would be reduced by $4\,p$.

And the entropy is now:

$$S_2 = \frac{S_1}{4\,p}$$

For fixing ideas, let us use the following values:

$$n = 1000$$
$$p = 50.$$

We would have for the first value of entropy (in J):

$S1 := evalf(subs(k_B = 1.38e - 23, n = 1000, m = 50, p = 50, k_B \cdot \ln(4\,n!)));$

8.16×10^{-20}

And for the second one:

$S2 := evalf\left(subs\left(k_B = 1.38e - 23, n = 1000, m = 50, p = 50, \frac{k_B \cdot \ln(4\,n!)}{4 \cdot p}\right)\right)$

4.08×10^{-22}

You see: the entropy is reduced, corresponding to more order. Now, the entropic contribution to the variation in Gibbs energy can be estimated, at room temperature:

$$eval(-300 \cdot (S2 - S1))$$

2.44×10^{-17}

This value (in J) is very small, but remember that we calculated for only one molecule. To compare it with the ΔH contribution that be estimated before ($3 \text{ kJ} \cdot \text{mol}^{-1}$), we have to divide the previous value by the Avogadro number:

$$\frac{5e3}{6e23} = 5 \times 10^{-21}$$

which is smaller.

8.1.4.3 Validation

In this case, we invite the students to read a much more thorough analysis of this problem (Silverstein, 1998; Biedermann *et al.*, 2014).

8.1.5 Conclusions and Perspectives

This exercise is primarily a way to remind us that "energy" (Gibbs energy) includes two main factors: enthalpy H and an entropic term $-TS$. In order to interpret phenomena (why salt melts ice on roads in winter, etc.), one needs to frequently calculate the entropic term. Certainly, Silverstein (1998) discusses other models that are not the one that we considered here, but the same kind of treatments could be applied as well. Would you like to do it? If so, remember that our models are simplistic and that specialists discuss the questions more thoroughly (Conti Nabali *et al.*, 2020).

8.2 WHY DO OIL DROPLETS MOVE UPWARD IN MAYONNAISE SAUCE?

8.2.1 An Experiment to Understand the Question

8.2.1.1 First, the General Idea

Many chefs say (and write) that you can "stabilize" a mayonnaise by the addition of hot vinegar to it when the emulsion is made. Is this true?

Certainly, we could trust starred chefs, *i.e.*, people that are considered excellent professionals, but let us remember that these people, albeit being artists, that is, being able to produce "beauty to eat", have often very little knowledge of chemistry and physics (it is not a reproach that I make to them). As a consequence, we should probably be cautious about the technical information given for cooking. Here, you can test this yourself:

1. Place one egg yolk in a vessel;

2. add one tablespoon of vinegar, salt and pepper to taste;

3. add about one teaspoon of oil, and whip until you do not see the oil (it has been divided in droplets dispersed in the aqueous phase, and these droplets are too small to be seen with the naked eye);

4. add another teaspoon of oil, and whip as before; repeat until the consistency becomes "thick";

5. divide the mayonnaise into five bowls:

   - one that you keep untouched, as a control,
   - one with a tablespoon of cold water: this will tell you if —in the assumption of a real effect of hot vinegar— water is the reason of the effect,
   - one with hot water: the difference with the previous sample will tell you if the temperature of water is important,

- one with one tablespoon of vinegar: here, the difference with the water will tell you a possible effect of acidity,

- one with a tablespoon of hot vinegar, as said by some chefs;

6. store the five bowls in the fridge and look at them day after day, possibly with a microscope (you simply take a small sample that you put on a glass plate, without a cover slide in order to avoid changing the structure).

8.2.1.2 Questions of Methods

The previous experiment is interesting, but can we improve it? Let us analyze that we look for the "stability" of the emulsion, for example against disruption: oil tends to move upward (to "cream") and accumulate at the surface, sometimes coalescing, and the aqueous liquid would flow down, as for a whipped egg white (do the experiment: after one night, a whipped egg white gives back a liquid egg white, that—contrary to what is said and written—you can whip again to make a whipped egg white).

Indeed, if you want to measure the thickness of layers of oil, it is more convenient to store the five emulsions in a test tube: a certain volume of oil will make a thicker layer, hence reducing the relative uncertainties.

I cannot resist telling you, even if it is a bit off topic, that this kind of experimental idea was taken to a wonderful extreme by Antoine Laurent de Lavoisier (1743–1794), the French chemist who created modern chemistry, putting this science on a quantitative basis, using precision tools (scale, among others). In particular, Lavoisier studied meat stock, and —in an article whose importance was neglected for a long time— he realized perfectly that the interest of stocks lies in the dry matter that it contains (Lavoisier, 1783; This, 2006). In order to measure this dry matter rapidly, he used a special densimeter that he created, with an empty glass bulb, on top of which he added a very thin rod, made by folding a sheet made of silver: in this way, a very small variation of density of the liquid (an increase of the density made the densimeter float up) was transformed into a large variation in height.

8.2.1.3 Where the Beauty Lies

About Lavoisier's method for measuring density, there is even more than the smartness of the densimeter: Lavoisier introduced a wonderful scientific "method of zero", or "opposition method". The goal being to measure tiny variations of x, you subtract a constant value almost equal to what you measure, so that you see only a small difference, that you can approach with more precision. For example, imagine that you want to determine small (let us say 1 A) intensity variations of an electric current of intensity close to 10^6 A: it would be very difficult to measure such variations because they are one millionth of the signal. However, if you subtract a constant electric current of exactly 10^6 A, then you will have to measure only a 1 A a variation of intensity close to 1 A, which is easy.

Coming back to Lavoisier, the thin rod that he used was allowing large variations of height, depending on the density of the liquid where the densimeter was put, but he had also invented putting a small cup on top of the silver rod, and adding very small weights (that you can measure very precisely, using a scale). In adding weights he was always putting the densimeter at the same level in the liquid, reducing the uncertainties.

Finally, we have to add that Lavoisier's method could be improved by another very important idea: keep in mind that you generally get more precision in measurements if you determine a frequency. More precisely, the idea is to look for a resonance, and this is why resonant circuits (remember your courses on RLC circuits, in electricity) are so widely taught at university: if the frequency that you measure is close to resonance, then the "signal" will be important. Otherwise, it will be weak (This, 2016a).

8.2.2 The Question

Many culinary systems are of colloidal nature, *i.e.*, they are made of particles of some phases (gas, liquid, solid) dispersed into continuous phase (gas, solid, liquid) (This, 2016b; IUPAC, 2019b). For example, "whipped egg whites" are foams, made of air bubbles dispersed in the viscous environment of the 10% protein solution of egg white. Many sauces, like béarnaise and hollandaise, are dispersions of both "oil" droplets and protein aggregates in a liquid phase, the latter coming for egg yolk (50%) or from vinegar (more than 90%): you make these sauces by whipping egg yolk with vinegar or white wine, and then you cook with butter. As the density of oil, air or solid particles is generally different from the density of the liquid continuous phase, such dispersed systems are not stable, thermodynamically speaking, and sedimentation or creaming can occur. In other words, in one case, particles (solid particles) fall, but in the other case (air bubbles, oil droplets...), they move up to the upper part of the sauce.

How fast does it occur? And does it occur always? The answers to such questions is important for science and technology education, because they allow us to describe common systems and also because they introduce scientific developments that can be useful parts of scientific culture.

Let us start with raw milk: how long should we wait in order to recover the cream? Or consider a mayonnaise sauce: how long can we keep it until the emulsion is disrupted?

Often a microscopic picture of the system is helpful. For example, Figure 8.5 shows a microscopic picture of a mayonnaise when about 40% of the volume is oil.

With more oil, the picture is different (Figure 8.6).

We want to calculate how objects can sediment or cream (the same question) in a liquid.

8.2.3 Analysis of the Question

8.2.3.1 The Data is Introduced

No data is given for this exercise.

8.2.3.2 Qualitative Model

For considering sedimentation and creaming, we build a model with one particle only in a liquid (*e.g.*, an oil droplet), assuming first that there are no interactions between

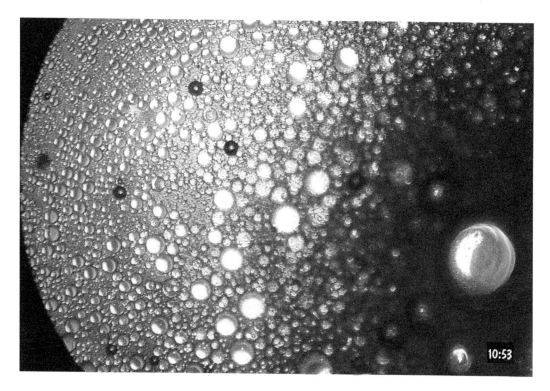

FIGURE 8.5 *Oil was whipped using a whisk in a 50:50 mixture of egg yolk and vinegar. The largest oil droplets have a diameter of 0.1 mm.*

neighboring particles in an emulsion with many such oil droplets. It can be observed that this assumption holds for diluted sauces, such as mayonnaise at the beginning of the production, but not for packed emulsions like ready-to-use mayonnaises (which are even sometimes considered as gels by some rheologists).

Then, we can assume that the particle (the oil droplet) that we consider is much bigger than the molecules of the liquid, so that the liquid seems to be "continuous" (an assumption that we have to keep in mind for later) (Figure 8.7).

A good practice is to "simplify" the problem, dropping all secondary aspects of it (which is why it is important to be able to distinguish different orders of magnitude). For example, here, the number of dimensions was reduced from three to two. One can go further, assuming a spherical particle (Figure 8.8).

In this picture the particle was put in the center of the liquid, on a vertical axis of symmetry, purely for aesthetics.

8.2.3.3 Quantitative Model

When the picture associated with a qualitative model is made, the problem can be translated into calculation, introducing formal variables: to this end, it is sufficient to describe what is seen; *i.e.*, a particle having a shape and a defined "nature" (shown by the color, on the picture), and a liquid, having a different nature (another color).

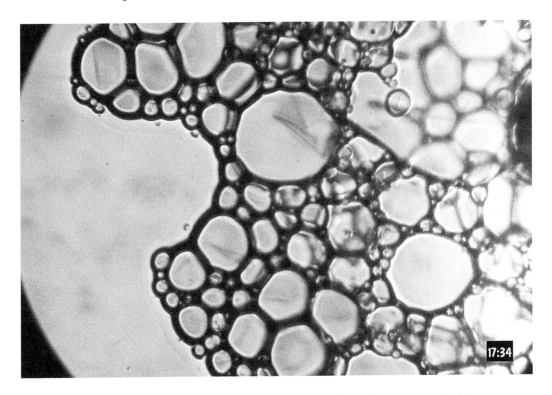

FIGURE 8.6 *Mayonnaise whipped with a whisk when the oil content is 80%.*

FIGURE 8.7 *One particle in a liquid. We now envision the successive steps of modeling. Of course, an oil droplet would not have this shape, but it is a good practice to be first very general, and then to make explicit assumptions.*

For the shape of the particle, one variable is enough: the radius r of the particle.
For its material, we shall use the density of the particle ρ_p.
For the liquid, we need both the density of the liquid ρ_l, and the viscosity η.
For the liquid, we do not consider its quantity, because the motion of the particle is the same, be a small or a large quantity of liquid in which the particle is moving. On the other hand, the viscosity is introduced because if there are motions in a liquid,

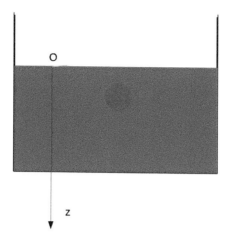

FIGURE 8.8 *Modeling the particle in the liquid. Assumptions are made.*

the liquid has to move away in front of the solid, whereas it "sticks" behind the solid (if it did not stick, vacuum would be created, and this would be "cavitation"; the nil pressure would draw rapidly the fluid toward the surface of the moving solid, except for very high velocities, when cavitation would occur). As a whole, the particle draws with it some liquid, and this is characterized by viscosity. Here, it has to be added that liquid flows are neglected, which is a very strong assumption in some cases, such as when a solid is sedimenting in a liquid inside a vessel whose radius becomes comparable to the flow disturbance.

Finally, as there is a motion, and because we assume this motion in one direction, in a particular sense, we have to introduce a vertical axis, toward the bottom, for example (Oz).

8.2.4 Solving

8.2.4.1 *Looking for a Solving Strategy*

Variables being introduced, relationships between them can be considered. Because we want to describe a sedimentation or a creaming, *i.e.*, a motion, we have to consider the equations of dynamics. The first equation, the "fundamental equation of motion", or "Newton's equation", indicates that the sum of forces acting on a body is equal to the product of the mass of this body by its acceleration. This leads us very logically to observe that we did not introduce the mass m of the particle, and its velocity v. We will certainly need these two variables.

8.2.4.2 *Implementing the Strategy*

Which forces act on the particle?

1. First the weight:

$$P = mg$$

2. Then Archimedes force (buoyancy), equal to the weight of the volume of liquid displaced by the body:

$$F = V \cdot \rho_l \cdot g$$

3. Finally, when the particle moves, there are frictions, which can be described using the "Stokes force" within the frame of the hypotheses done before:

$$S = 6 \cdot \pi \cdot \eta \cdot v \cdot r$$

Now, some new variables appear, such as the velocity v (we can write $v(t)$ if it help us to remember that this velocity is changing with time). One can try to link them to variables already introduced: for example, to the velocity:

$$v(t) = \frac{d}{dt} z(t);$$

So that the fundamental equation of motion, assuming a sedimentation, is:

$$eq1 : \ P - F - S = m \cdot \frac{d}{dt} v(t)$$

Let us observe now that there is a relationship between the mass and the volume of the particle:

$$m = \rho_p \cdot V$$

And this means a relationship between the mass and the radius:

$$V = \frac{4}{3} \cdot \pi \cdot r^3$$

Finally, the equation of the motion becomes:

$$\frac{4}{3}\rho_p \pi r^3 g - \frac{4}{3}\pi r^3 \rho_l g - 6\pi\eta \left(\frac{d}{dt} z(t) \right) r = \frac{4}{3}\rho_p \pi r^3 \left(\frac{d^2}{dt^2} z(t) \right)$$

In this equation, we see the first and second derivative of the function $z(t)$: this means that it is a "differential equation of the second order" (the highest order derivative is the second one); and because there are no products of the function and its derivatives (such as $z(t)$, or $\frac{d}{dt}(z(t)) \cdot \frac{d^2}{dt^2}(z(t))$), it is "linear". In the past—even the recent one—, one would have performed some mathematics in order to solve the differential equation, but using a software for formal calculation such as *Maple*, it is very simple and involves using the *dsolve* command (solving differential equations, which is different from the *solve* command for ordinary equations or systems of equations).

Here, the simplest way to do this is straightforward: you simply copy and paste the differential equation inside the *dsolve* command:

$$dsolve \left(\frac{4}{3}\rho_p \pi r^3 g - \frac{4}{3}\pi r^3 \rho_l g - 6\pi\eta \left(\frac{d}{dt} z(t) \right) r = \frac{4}{3}\rho_p \pi r^3 \left(\frac{d^2}{dt^2} z(t) \right) \right)$$

And when you press "enter", you get:

$$z(t) = -\frac{2\rho_p r^2 e^{-\frac{9\eta t}{2\rho_p r^2}} \_C1}{9\eta} - \frac{2g\, r^2\,(\rho_l - \rho_p)\,t}{9\eta} + \_C2$$

The result given includes three terms and two constants "$\_C1$" and "$\_C2$". In order to find them, we have to consider the initial conditions.

The are two possibilities. The first one would be to calculate the velocity, as the derivative of $z(t)$, and writing that it is equal to 0 at $t = 0$; also we would have to write that the initial starting point is 0 at this time. The second one is to introduce directly the initial conditions in the dsolve command. Here we shall use both, one to obtain the result, and the second to validate.

Let us begin by calculating the derivative of $z(t)$: with the software, it is very simple, because you simply copy and paste the expression of $z(t)$ in the $\frac{d}{dx}(f)$ operator:

$$\frac{d}{dt}\left(-\frac{2\,r^2\rho_p e^{-\frac{9}{2}\frac{\eta t}{r^2\rho_p}}\_C1}{9\,\eta} - \frac{2\,r^2 g(-\rho_p + \rho_l)t}{9\,\eta} + \_C2 \right)$$

$$e^{-\frac{9}{2}\frac{\eta t}{r^2\rho_p}}\_C1 - \frac{2\,r^2 g\,(-\rho_p + \rho_l)}{9\,\eta}$$

Now, we solve the equation:

$$solve\left(e^{-\frac{9}{2}\frac{\eta t}{r^2\nu_p}}\_C1 - \frac{2\,r^2 g(-\rho_p + \rho_l)}{9\,\eta} = 0, \_C1 \right)$$

$$\frac{2\,r^2 g\,(-\rho_p + \rho_l)}{9\,e^{-\frac{9}{2}\frac{\eta t}{r^2\rho_p}}\eta}$$

This result holds for $t = 0$:

$$\frac{2\,r^2 g(-\rho_p + \rho_l)}{9\,e^{-\frac{9}{2}\frac{\eta t}{r^2\rho_p}}\eta} \quad \xrightarrow{\text{evaluate at point}} \quad \frac{2\,r^2 g\,(-\rho_p + \rho_l)}{9\,\eta}$$

We introduce this value in the general equation:

$$z(t) = -\frac{2\,r^2\rho_p e^{-\frac{9}{2}\frac{\eta t}{r^2\rho_p}}\cdot\left(\frac{2\,r^2 g(-\rho_p+\rho_l)}{9\,\eta}\right)}{9\,\eta} - \frac{2\,r^2 g(-\rho_p + \rho_l)t}{9\,\eta} + \_C2$$

$$z(t) = -\frac{4\,r^4\rho_p e^{-\frac{9}{2}\frac{\eta t}{r^2\rho_p}}g\,(-\rho_p + \rho_l)}{81\,\eta^2} - \frac{2\,r^2 g\,(-\rho_p + \rho_l)\,t}{9\,\eta} + \_C2$$

And we write that z(0) = 0, and solve for the second constant:

$$solve\left(0 = -\frac{4\,r^4\rho_p e^{-\frac{9}{2}\frac{\eta t}{r^2\rho_p}}g(-\rho_p + \rho_l)}{81\,\eta^2} - \frac{2\,r^2 g(-\rho_p + \rho_l)t}{9\,\eta} + \_C2, \_C2 \right)$$

$$\frac{2}{81}\frac{r^2 g\,(-\rho_p + \rho_l)\left(2r^2\rho_p e^{-\frac{9}{2}\frac{\eta t}{r^2\rho_p}} + 9\eta t\right)}{\eta^2}$$

Again, this result is for $t = 0$:

$$\_C2 = \frac{2}{81}\frac{r^2 g(-\rho_p + \rho_l)(2r^2 \rho_p)}{\eta^2}$$

$$\_C2 = \frac{4}{81}\frac{r^4 g(-\rho_p + \rho_l)\rho_p}{\eta^2}$$

Finally:

$$z(t) = -\frac{2}{9}\frac{r^2 \rho_p e^{-\frac{9}{2}\frac{\eta t}{r^2 \rho_p}} \cdot \left(\frac{2}{9}\frac{r^2 g(-\rho_p + \rho_l)}{\eta}\right)}{\eta} - \frac{2}{9}\frac{r^2 g(-\rho_p + \rho_l)t}{\eta} + \frac{4}{81}\frac{r^4 g(-\rho_p + \rho_l)\rho_p}{\eta^2}$$

$$z(t) = -\frac{4}{81}\frac{r^4 \rho_p e^{-\frac{9}{2}\frac{\eta t}{r^2 \rho_p}} g(-\rho_p + \rho_l)}{\eta^2} - \frac{2}{9}\frac{r^2 g(-\rho_p + \rho_l)t}{\eta} + \frac{4}{81}\frac{r^4 g(-\rho_p + \rho_l)\rho_p}{\eta^2}$$

Here it is... but is it right?

8.2.4.3 Validation

The first method, that we used above, is quite cumbersome; another possibility would be to introduce directly the initial conditions in the *dsolve* command. To this end, we first define the initial conditions:

$$restart; ics := z(0) = 0, \mathrm{D}(z)(0) = 0$$

$$z\left(0.00 \times 10^0\right) = 0.00 \times 10^0, \mathrm{D}(z)\left(0.00 \times 10^0\right) = 0.00 \times 10^0$$

And now, we solve the equation using them:

$$dsolve\left(\left\{\frac{4}{3}\rho_p \pi r^3 g - \frac{4}{3}\pi r^3 \rho_l g - 6\pi\eta\left(\frac{\mathrm{d}}{\mathrm{d}t}z(t)\right)r = \frac{4}{3}\rho_p \pi r^3\left(\frac{\mathrm{d}^2}{\mathrm{d}t^2}z(t)\right), ics\right\}, z(t)\right):$$

And we get the same result as before.

8.2.5 Expressing the Results

8.2.5.1 The Formal Result Found is Written Down Again

Again, we give the solution that was found:

$$z(t) = -\frac{4}{81}\frac{r^4 \rho_p e^{-\frac{9}{2}\frac{\eta t}{r^2 \rho_p}} g(-\rho_p + \rho_l)}{\eta^2} - \frac{2}{9}\frac{r^2 g(-\rho_p + \rho_l)t}{\eta} + \frac{4}{81}\frac{r^4 g(-\rho_p + \rho_l)\rho_p}{\eta^2}$$

8.2.5.2 Discussion

In the last equation, one can see three terms:

- a constant (the last term: $\frac{4}{81}\frac{r^4 g(-\rho_p + \rho_l)\rho_p}{\eta^2}$), which is quite uninteresting because it tells only at which level the particle was released at time $t = 0$ (with no velocity);

- a term including a negative exponential as a function of t ($e^{-\frac{9}{2}\frac{\eta t}{r^2\rho_p}}$): this term decreases very rapidly with time, so that it is important only at the onset of motion;

- a term proportional to t ($-\frac{2}{9}\frac{r^2 g(-\rho_p+\rho_l)t}{\eta}$), very preponderant after the initial acceleration; after the start, it becomes the main one.

Finally, we see more clearly that after the initial acceleration (rapidly decreasing toward 0), the velocity is almost at constant speed. The regime after the initial acceleration is called "stationary". It can be reached when the acceleration is nil:

$$\frac{4}{3}\rho_p\pi r^3 g - \frac{4}{3}\pi r^3 \rho_l g - 6\pi\eta\left(\frac{\mathrm{d}}{\mathrm{d}t}z(t)\right)r = 0$$

Here, we get only the last two terms, and the limit velocity is:

$$v_{\lim} = \frac{\mathrm{d}}{\mathrm{d}t}\left(-\frac{2}{9}\frac{r^2 g(-\rho_p+\rho_l)t}{\eta} + \_C1\right) = -\frac{2}{9}\frac{r^2 g(-\rho_p+\rho_l)}{\eta}$$

One can see that:

- the greater the difference in densities, the faster the sedimentation with a g factor (and this is why centrifugation is used, in laboratories and in the industry, in particular for cream separation);

- the greater the viscosity, the slower the sedimentation;

- the smallest radiuses are associated with slow sedimentation, but the variation is in r^2!

One can also observe that if one knows three of the four parameters (r, ρ_p, ρ_l, η) and if one measures the limit velocity, then the fourth can determined!

Finally, it is worth considering the question: how important is the limit velocity? Or when does the particle reach this limit velocity? *Stricto sensu*, never, but one can ask when it is 99% of it, or 99.9%, for example.

More generally, let us use a velocity which is a fraction of the limit velocity:

$$V = p \cdot v_{\lim}$$

We now ask when this velocity is obtained:

$$p \cdot \left(-\frac{2}{9}\frac{r^2 g\left(-\rho_p+\rho_l\right)}{\eta}\right)$$

$$= \frac{\mathrm{d}}{\mathrm{d}t}\left(-\frac{4}{81}\frac{r^4\rho_p e^{-\frac{9}{2}\frac{\eta t}{r^2\rho_p}}g\left(-\rho_p+\rho_l\right)}{\eta^2} - \frac{2}{9}\frac{r^2 g\left(-\rho_p+\rho_l\right)t}{\eta} + \frac{4}{81}\frac{r^4 g\left(-\rho_p+\rho_l\right)\rho_p}{\eta^2}\right)$$

The solution is:

$$t = -\frac{2}{9}\frac{\ln(-p+1)\,r^2\rho_p}{\eta}$$

Then, we explore this solution for spherical particles with a radius of 1 mm, of density four times the density of water, with viscosity $\eta = 10^{-3}$ kg/m/s. We find 4.09. Which units? The answer is seconds, because we always use SI units.

With a lower density of particles (1200 kg · m^{-3}), one would find 1.23 s.

What about a smaller radius for the particles? For example, for this second density, but with a radius ten times smaller, the time would be in 0,0123 s, which is faster, but there is something more: having done three calculations, we also see that this exploration of the result is not systematic. Students can now understand that they need to plot a "response surface", for example.

8.2.6 Why is the Stokes Equation in r, and not in r^2

The Stokes equation is in r, which can seem strange, because we know well that if we extend a finger outside a moving car, we perceive less force than if we extend the whole hand. In a word, the Stokes force "should" be in r^2 rather than in r.

We shall not have the entire discussion here, but because software are so easy to use, you can try to see what the result would be with other expressions for the drag force in r^2 or r^3.

However, remember that the expression $6\pi\eta vr$ corresponds to a force, in newtons. Changing r to r^2 or r^3 can be made only if you introduce a constant with the dimension of the inverse of a distance, or the invert of the square of a distance. Keeping an eye on dimensions is useful in physics!

8.2.7 Dipping in Microscopy

In the previous section, it was not explained how to get the solution of the differential equation from *Maple*, *i.e.*, a software for symbolic computation. Here, we want to stress that such types of software are key to student learning. We show that in the next step in exploring sedimentation and creaming for small particles.

In what was done previously, particles much bigger than water molecules were considered (for example, with particles of radius 1 mm, the diameter ratio is $\frac{10e-3}{1e-10}$ $= 1.0 \times 10^7$, which is very large). But water is indeed not a continuous medium, and when particles are very small, as for example with casein micelles in milk (diameter 150 nm), then the water molecules—which move with a kinetic energy of $\frac{3k_BT}{2}$—would appear in the picture.

If water molecules bumping against a stone don't cause it to move much, on the contrary they have a big effect on small and light particles: Brownian motion can be observed.

Now, science from the 18th century is not sufficient, and this is fine, as it pushes the students into the 19th century, with statistical thermodynamics (Reif, 1967).

Let us now imagine a vessel with two flat levels, or one step. One level will be called "up", and the other "down. Small particles (red) are bumped by water molecules, so that some of the particles of the up level can go down, and some of the particles of the down level can go up. Of course, as there is gravity, one could imagine that there are more particles on the bottom than on the top level (for particles having a density

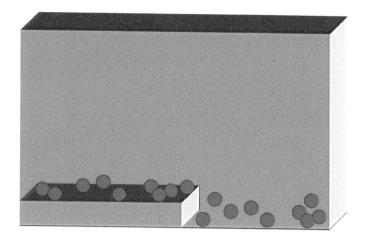

FIGURE 8.9 *A simple equilibrium for molecules in water. For simplicity, two levels can be considered.*

much greater than water), but how many? Thermodynamics considers equilibriums (Figure 8.9).

At the equilibrium, the "Maxwell-Boltzmann distribution" gives the proportion of particles of energy E, comparing this energy with the thermal motion energy $k_B \cdot T$, where k_B is the Boltzmann constant and T is the temperature (in kelvins, K).

Using the Maxwell-Boltzmann distribution, we would have the number of particles at a height h:

$$n(h) = K \exp\left(-\frac{\phi(h)}{k_B \cdot T}\right)$$

where $\phi(h)$ is the potential energy at height h, and K is a constant, given by:

$$\phi(h) = \int_0^h mg\left(1 - \frac{\rho_w}{\rho_p}\right) dh = mgh\left(1 - \frac{\rho_w}{\rho_p}\right)$$

Thus:

$$n(h) = n(0) \exp\left(-\frac{\left(mgh\left(1 - \frac{\rho_w}{\rho_p}\right)\right)}{k_B \cdot T}\right)$$

This is an exponential function: either the particles are mostly grouped at the surface, or they are at the bottom.

But remember that this distribution cannot apply for molecules, and hardly at all for small particles.

8.2.8 Conclusions and Perspectives

This exercise was discussed in more detail because we wanted to show that simple knowledge is important for the initial discussion of common phenomena in the kitchen, but we also wanted to show the limits of such simple treatments.

8.3 HOW MANY SURFACTANT MOLECULES ARE THERE AT THE OIL-WATER INTERFACE IN MAYONNAISE, AND WHAT IS THE MASS OF SURFACTANT NEEDED TO MAKE THE SAUCE?

8.3.1 An Experiment to Understand the Question

8.3.1.1 First, the General Idea

Here, we shall explore experimentally a very popular appetizer in France: a hard boiled egg with mayonnaise. Yes, again mayonnaise... because this sauce is said to be a state religion in France, as wrote the American writer Ambrose Bierce in his *Devil's Dictionary* (Bierce, 1911). For such a dish, of course, the egg should be properly cooked (no green color around the yolk, no sulfurous odor, etc.) and the sauce should be correctly emulsified; however, the question that we study here is more difficult: how to make a hard boiled egg with mayonnaise from only one egg?

In order to prepare the answer, let us make the following experiment:

1. Put an egg yolk in a vessel, and add one tablespoon of vinegar, and salt, and pepper;

2. whip while adding oil, teaspoon by teaspoon until a firm emulsion is done;

3. then, divide the sauce in two parts;

4. continue adding oil and whipping the first half: sooner or later, the sauce will fail;

5. add two tablespoons of vinegar to the second half, and add oil while whipping until the sauce is firm;

6. then, add vinegar again so that the consistency is decreased and add oil again while whipping hardly: this shows that at some point, when the sauce becomes viscous, there is enough water to pack more oil droplets in it.

Which volume of oil can be added if you have enough water?

8.3.1.2 Questions of Methods

The previous experiment is qualitative and not quantitative. How to turn it into something better? Certainly, the use of a scale is important, as said earlier, because mass measurements are much more precise than volume measurements.

Also, in the assumption that one egg yolk could make a large quantity of mayonnaise, using a whole egg as we did is a bad option because it would use a lot of oil. It is much better to use a small quantity.

Indeed, this last remark is important, because it has echoes in practical sessions in high schools or universities: too often, during practical sessions in chemistry, students are invited to experiment with quantities as big as 1 g, and this should be absolutely avoided, knowing that research laboratories try to use as few material as possible. Yes, in the past, when scientists wanted to extract hormones, for example, they had to boil tons of urine of pregnant mares in order to get micrograms of final products;

e.g., hormones. But there are many arguments in favor of reducing the quantities used in chemistry:

- with less product, there is less danger: for example, milligrams that explode would perhaps break a glass, but not a building, killing people;

- with less products, reduced quantities of solvents can be used (and these solvents, being organic ones, are not without risk);

- less product means less costs (for some of the chemicals used in research, the price can be thousands of dollars for milligrams);

- less products means less effluents.

Then, always think microchemistry!

8.3.1.3 *Where the Beauty Lies*

Let us give a recipe to finish this experimental part, answering our initial question: How to make a hard boiled egg with mayonnaise from only one egg?

1. Using a syringe, pierce the shell of an egg and dip the tip of the syringe into the yolk; suck out one drop of yolk;

2. boil the egg, for 10 minutes;

3. put the drop of yolk in a bowl with a teaspoon of vinegar and whip, adding tiny quantities of oil at a time; when the sauce becomes thick, add vinegar, and continue this way until you make about 100 g of sauce;

4. add salt and pepper to taste;

5. cut the hard boiled egg in two, and serve it along with the mayonnaise and possibly some leaves of salad, black olives, slides of tomato, some crushed garlic, preserved anchovies, dice draw (peeled) tomatoes, or strips of sun-dried tomatoes, ciseled shallots, chilies, and Worcestershire sauce.

8.3.2 The Question

Remember that the question of the section was not the one that we explored experimentally first. We wanted initially to calculate how many surfactant molecules there are at the oil-water interface in a mayonnaise, and what is the mass of surfactant that is needed to make the sauce. Put more clearly: let us calculate the minimum mass of surfactant and of water needed to make a volume V of mayonnaise.

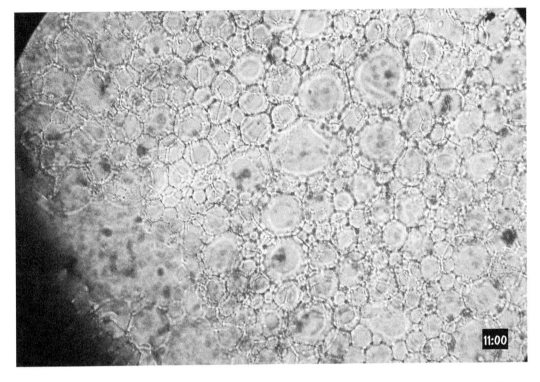

FIGURE 8.10 *Oil droplets are deformed when the oil proportion is high, in an emulsion such as mayonnaise.*

8.3.3 Analysis of the Question

8.3.3.1 The Data is Introduced

No data is given: one has to introduce it. Of course, in all the previous chapters we had to do such introduction often, but it is probably useful to repeat here that for such a calculation one good practice is to introduce a letter each time one quantity is to be described, let this quantity be known or unknown. This will be done in the "Quantitative Model" section below.

8.3.3.2 Qualitative Model

When a mayonnaise is made, it is often described as a dispersion of spherical oil droplets in water, but this is not true: in the finished sauce, the oil droplets are so closely packed that they are polyhedral (Figure 8.10).

8.3.3.3 Quantitative Model

Let us consider an emulsion of total volume V. The deformed droplets are assumed to be cubes, with (at the limit) one monomolecular water layer between the cubes, surfactants being at the oil/water interface, on both sides of the water layer.

Of course, the issue of assuming cubes instead of more complex polyhedra can be discussed, as well as assuming only one "layer" of surfactants at each oil/water interface. More generally, it is safe to keep explicit track of all assumptions made during a calculation, and to come back to them in the "discussion" section.

8.3.4 Solving

8.3.4.1 Looking for a Solving Strategy

Simply moving from macroscopic to microscopy will lead to the solution.

8.3.4.2 Implementing the Strategy

For each "oil cube", the size of a side is c. So that the volume is $v = c^3$.

Each cube has six faces, so that the area of the surface of one cube is:

$$a = 6c^2$$

If we neglect the thickness of the water films (remember the calculation made in Section 7.3), using V, the volume of oil, the number of droplets is such as:

$$N \cdot c^3 = V$$

So that the total area A of the oil/water interface is N times the area of a cube:

$$A = Na = \frac{V}{c^3} \cdot 6c^2 = \frac{6V}{c}$$

Let now σ be the area of a surfactant molecule (phospholipid, protein, etc.). The number n of molecules of surfactant necessary to cover this surface is equal to the total area by the area of one molecule:

$$n = \frac{A}{\sigma} = \frac{6V}{c\sigma}$$

From this number of molecules, we deduce the number of moles of surfactant needed:

$$n_m = \frac{6V}{c\sigma N_A}$$

with N_A the Avogadro number.

And the mass of surfactant is found using the molar mass M of the surfactant:

$$m = n_m M = \frac{6VM}{c\sigma N_A}.$$

Finally, it is easy to calculate as well the minimum number of water molecules assuming their area is s:

$$n_e = \frac{A}{s} = \frac{6V}{c \cdot s}$$

And their mass:

$$m_e = \frac{6VM_e}{csN_A}$$

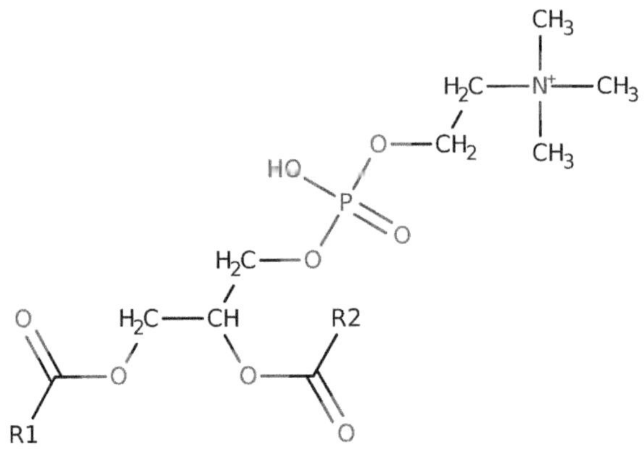

FIGURE 8.11 *Phosphatidylcholines are phospholipids; more precisely, they are cholic esters of phosphatidic acids.*

8.3.5 Numerical Application

We assume 1 L of emulsion $(10^{-3}m^3)$, and droplets with side equal to 10^{-5} m.
For the surface, we consider two cases:

1. For phospholipids: remember that the two hydrocarbon tails are in oil, whereas the head is in water, because of its electric charge (Figure 8.11). Without taking into account the electric repulsion, the surface would be about the square of two bonds only (for the head); however, one has to remember that the globular volumes shown in popular or educational pictures consider the électric repulsion, and this is why it is best to rely on measurements and to consider a surface of 0.70 nm² (NCBI, 2002).

2. For proteins: let us considered them as coiled in a plane, the amino acid residues being on the edges of a rectangular grid of size $p \times p$. With 450 amino acid residues, the size of this square would be

$$p = evalf(\sqrt{450}) = 21.21.$$

Assuming 3 covalent bonds per residue of amino acid, the area would be (we use the value 0.15 nm for a carbon-carbon bond):

$$(21.21 \cdot 3 \cdot 1.5e - 9)^2 = 9.11 \times 10^{-15} m^2.$$

For the number of molecules :

- for phospholipids:

$$subs\left(V = 1e-3, c = 1e-5, \sigma = 0.7e-18, n = \frac{6\,V}{c\,\sigma}\right) n = 8.57 \times 10^{20}$$

- for proteins:

$$subs\left(V = 1e-3, c = 1e-5, s = 9.11e-15, n = \frac{6\,V}{c\,s}\right) n = 6.59 \times 10^{16}$$

For the mass needed:

- of phospholipids (in S.I, *i.e.*, in kg; here we use the molar mass of 1-stearoyl-2-linoleoyl-sn-glycero-3-phosphatidylcholine):

$$subs\left(V = 1e-3, c = 1e-5, \sigma = 0.7e-18, M = 0.800, N_A = 6e23, \frac{6\,V\,M}{c\,\sigma\,N_A}\right)$$

$$1.14 \times 10^{-3}$$

- of proteins (in kg):

$$subs\left(V = 1e-3, c = 1e-5, s = 9e-15, M = 45, N_A = 6e23, \frac{6\,V\,M}{c\,s\,N_A}\right)$$

$$5.00 \times 10^{-6}$$

8.3.6 Conclusions and Perspectives

Let us draw first the conclusion of the previous calculations: with our assumptions, for making 1 L of mayonnaise, one would need only 1 g of phospholipids, and 5 mg of proteins. The quantities being ordinarily used for cooking are much higher than what is calculated here.

Of course, the system that we considered was extreme, and we can try to test experimentally the result. It is not difficult to get an emulsion with only one drop of egg yolk and vinegar (water can be the limiting factor).

8.4 WHY ARE PASTIS, OUZO, OR ARAK TURBID ONLY AT THE SURFACE WHEN WATER IS ADDED SLOWLY, BUT BECOME TURBID IN THE ENTIRE GLASS AFTER SOME TIME?

8.4.1 An Experiment to Understand the Question

8.4.1.1 First, the General Idea

Here, we want to reproduce the turbidity that one can observe in drinks made from Pastis (in France), or Ouzo (in Greece), or Arak (Levantine). These beverages are made of water, ethanol, and an anise flavor due to anethole (Figure 8.12).

Let us first observe that, in spite of the -ole ending of its name, anethole is not an alcohol. It is poorly soluble in water, but it is soluble in ethanol. And it remains soluble in an ethanol-water mixture when the ethanol proportion is high (about 40%

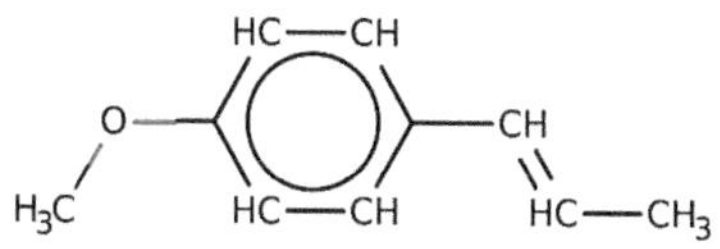

FIGURE 8.12 *Anethole, or 1-methoxy-4-[(1E)-prop-1-en-1-yl]benzene, is a major compound from the essential oil of anis.*

FIGURE 8.13 *When Pastis is slowly added to water, only the upper layer is turbid.*

in the drinks commercially sold). But when the beverage is added to water, anethole separates from the solvent, and makes dispersed droplets in the drink.

In order to reproduce this phenomenon of turbidity production:

1. prepare a glass of water;

2. in a test tube, put about less than 1 cm^3 of ordinary oil;

3. in this tube, add 5–10 cm^3 of ethanol or of a brandy (such as vodka);

4. shake vigorously the tube, and pour its content slowly into water, from a small height: a turbid layer is produced towards the surface (Figure 8.13);

5. store the glass, and look at the turbidity the next day: you see that it moved downward.

8.4.1.2 Questions of Methods

Why the turbidity? Indeed, oil was not entirely dissolved in ethanol, but it was transferred under a dispersed form in water, where it aggregated and made droplets. These droplets are responsible for the turbidity: remember that white light reflects on their surface.

More precisely, when light falls on an object, it excites the electrons and brings them into oscillation so that they radiate. For a perfectly homogeneous medium without local variation of the refractive index on the scale of the wavelength of light, the secondary waves interfere destructively except in the original direction of the light: the medium appears transparent, and the incident beam is only visible by looking in the forward direction. Now, if the medium has heterogeneities with typical sizes of the order of the light wavelength, the radiation is scattered into all directions and the medium appears cloudy. When investigating the turbidity of pastis, Isabelle Grillo and her colleagues (Grillo, 2003) did not use light, but rather neutron beams, and they discovered that a glass of Pastis scatters the light because it contains heterogeneities with a typical size close to the light wavelength.

Complementary to light scattering, small-angle neutron scattering (SANS) is a powerful tool to explore the shape and size of heterogeneities in matter with typical sizes from a few tens of nanometers up to hundreds of nanometers. Neutrons are sensitive to the nucleus of the atoms and a unique feature is the possibility of molecular labeling or contrast variation by replacing light water by heavy water, for example. In this way, radiuses for droplets between 0.2 and 0.8 micrometers were detected in pastis.

8.4.1.3 *Where the Beauty Lies*

Decades ago, I proposed to use the "Pastis effect" in the kitchen (This, 2003). In particular, I explained to Pierre Gagnaire, one of the very best chefs of the world, how to use it, and he devised a recipe that I translate now so that it is easier to make:

1. in a vessel, mix oil, calvados (an apple brandy) and an apple cut into four pieces; store in the fridge for one day;

2. take out the leaves of an artichoke, and cook the heart plus the leaves in a pan with cold and salted water (boil for 30 min for the heart, plus an additional 30 min for the leaves);

3. take out the artichoke, and filter the liquid;

4. add the content of the vessel to the liquid and infuse for 10 min;

5. put a filet of cod in a dish with butter and salt; cook for 20 min in the oven at 80°C;

6. distribute the fish in four soup plates, with the heart of artichoke diced and divided;

7. bring the stock to a boil, add pepper and salt to taste, and pour it in the plates.

   You see: we are now far away from our first experiment but the idea remains.

8.4.2 The Question

If you pour Pastis in water cautiously, only the upper layer is turbid at first, as in Figure 8.13; however, after some time, the entire system becomes turbid. The same effect occurs when a mixture of ordinary oil (sunflower, for example) and ethanol is poured on water. How can we explain that the less dense oil (possibly with ethanol) droplets go down in water?

8.4.3   Analysis of the Question

8.4.3.1   The Data is Introduced

No data is given.

8.4.3.2   Qualitative Model

The downward migration of oil droplets over time can appear paradoxical, because the density of oil and ethanol is less than the density of water, but this migration is a fact!

Here, the simple description that we used for the creaming of mayonnaise is no longer valid, and we have certainly to look more in depth. And if we do this, we recognize that the droplets are so small that there is a possibility that Brownian motion moves the droplets in all directions, so that they distribute throughout the entire glass.

8.4.3.3   Quantitative Model

We simply assume oil droplets of radius r.

8.4.4   Solving

8.4.4.1   Looking for a Solving Strategy

For this exercise, we remember the Maxwell-Boltzmann distribution that was used before in conditions where it should not have been (see Section 6.7). However, we said that the distribution was giving the same results as the differential equation used by Jean Perrin, Nobel prize in physics, for describing colloids (Perrin, 1926).

8.4.4.2   Implementing the Strategy

Using the Maxwell-Boltzmann distribution, the proportion of droplets at height h would be given by:

$$n(h) = K \exp(-\frac{mgh}{k_B T})$$

The equation includes an exponential in which we have to compare the numerator (gravity forces) and the denominator (kinetic energy of water).

When they are of the same order of magnitude, the mass can be found:

$$m = \frac{k_B T}{gh}$$

We also know:

$$m = \frac{\varrho 4\pi r^3}{3}$$

with where ϱ is the density and r is the radius.
So that we can determine the radius:

use $RealDomain$ **in** $solve$ $\left(\dfrac{\rho \cdot 4\pi r^3}{3} = \dfrac{k_B T}{gh}, r\right)$ **end use** $\dfrac{1}{2}\dfrac{6^{1/3}\left(Tk_Bg^2h^2\rho^2\right)^{1/3}}{\pi^{1/3}gh\rho}$

8.4.5 Expressing the Results

8.4.5.1 The Formal Result Found is Written Down Again

$$r = \frac{1}{2}\frac{6^{1/3}\left(Tk_Bg^2h^2\rho^2\right)^{1/3}}{\pi^{1/3}gh\rho}$$

8.4.5.2 Finding Digital Data

All data can be found in previous exercises.

8.4.5.3 Introduction of Data in the Formal Solution

$evalf\left(subs\left(\rho = 800, k_B = 1.38e-23, g = 9.8, h = 1, T = 300, \dfrac{1}{2}\dfrac{6^{1/3}\left(k_BT\rho^2\pi^2g^2h^2\right)^{1/3}}{\rho\pi gh}\right)\right)$

5.01×10^{-9}

This would be only one order of magnitude larger than the size of anethole droplets detected in Pastis.

8.4.6 Conclusions and Perspectives

In our calculation, we found the radius for which the two main sources of energy are equal, but of course, sedimentation and creaming can occur for smaller and for bigger droplets.

We can also use the characteristic length that was found in Section 6.7:

$$L = -\frac{k_B \cdot T}{m \cdot g \cdot \left(1 - \frac{\varrho_w}{\varrho_p}\right)}$$

For the Pastis effect, an order of magnitude of L is:

$subs\left(k_B = 1.38e-23, T = 300, m = \dfrac{1000 \cdot 4 \cdot (1e-7)^3 \cdot 3.14}{3},\right.$

$\left. g = 9.8, \varrho_w = 1000, \varrho_p = 800, -\dfrac{k_B \cdot T}{m \cdot g \cdot \left(1 - \frac{\varrho_w}{\varrho_p}\right)}\right)$

4.04×10^{-4}

This value (about 0.4 mm) is certainly compatible with a Brownian motion, and a Maxwell-Boltzmann distribution of oil droplets.

This means that the structures responsible for turbidity are subject to Brownian motion: the can move in any direction, in particular toward the bottom.

8.5   IS BUTTER AN EMULSION?

8.5.1   An Experiment to Understand the Question

8.5.1.1   First, the General Idea

In order to investigate the composition of butter, let us begin by heating very slowly some of it in a transparent vessel (Pyrex glass on a stove, at low heat, for example): when butter melts, you can observe the formation of a milky layer at the bottom, with, above, a clear, transparent, yellow layer, and some scum at the upper surface. Indeed, butter is mainly made of fat, and up to 18% water: when the fat melts, the aqueous solution (water plus lactose and proteins) is released from the matrix and falls to the bottom of the vessel, leaving a layer of pure fat (triglycerides) that is called "clarified butter".

Of course, this separation of butter into fat and aqueous solution can be reversed, using the ordinary process for making emulsions:

1. place in a vessel the aqueous solution recovered after clarification of butter;

2. add the cooled clarified butter drop by drop, while whipping.

   A creamy material like butter is obtained. If you store it in the fridge, you obtain something like butter.

Is it an "emulsion"? It is the question studied here, and we need another experiment:

1. take the clarified butter and store it in a room at a temperature of about 30°C;

2. when you see some solids formed in the liquid, try to scoop them out;

3. now put the remaining liquid in a room at a cooler temperature or in the fridge and wait until solids form; again try to separate them;

4. if possible, make another separation at 0°C.

Finally, this process of "fractional crystallization" shows that butter fat is made of various fractions.

8.5.1.2   Questions of Methods

Let us begin simply: the fat in butter is mainly (98%) made of triglycerides (Lopez *et al.*, 2006), There are many kinds of triglycerides in milk (and butter) fat, as the number of possible fatty acid residues is about 400 (Lopez *et al.*, 2013). This explains why the melting curve of fat is smooth, instead of being very sharp, as is it often for pure compounds such as water (Figure 8.14).

Let us elaborate on the last sentence corresponding to ice melting. Since water is made of only one kind of molecule, the phase transition occurs abruptly. But for a mixture of two compounds in equal quantity, assuming that they would be

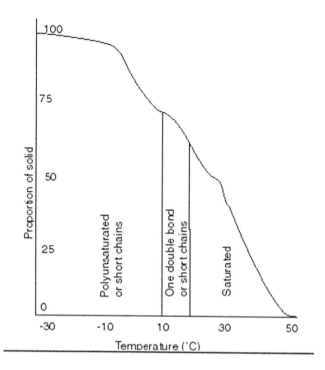

FIGURE 8.14 *The proportion of solid in butter as a function of temperature.*

independent (plus other conditions), the proportion of solid would decrease first by one half, and then it would drop to zero when the second transition was reached. Here, for butter, the curve is obviously the result of the successive melting of the various triglycerides, by order of melting point but with complexity arising from the fact that triglycerides exhibit "allotropy"; *i.e.,* they can make various kind of solids, as chocolate makers knows well and the reason why they have to "temper" their melts.

Indeed, in cocoa butter, the triglycerides have mainly only three fatty acid residues, which are of oleic acid (O, 18 carbon atoms), stearic acid (S, 18 carbon atoms, but differently organized) and palmitic acid (P, 16 carbon atoms). For most triglycerides of the cocoa butter, the oleic acid residue is in the middle position on the glycerol residue. This structure therefore essentially determines the crystallization behavior of the cocoa butter.

Six different polymorphs are formed after crystallization of cocoa butter. Crystals in the phase γ (I) appears when the molten chocolate is cooled rapidly, and it is responsible of the white appearance of chocolate ("fat bloom"). The melting point of the phase α (II) is 23.3°C; crystals of this kind form for cooling rates of 2°C per minute; they are more stable than the phase γ (I). Phase β' (III) and phase β' (IV) are similar, with melting temperatures of 24.5°C and 27.3°C. Crystals of these phases can be produced with cooling rates of 5°C to 10°C per minute. Phase β (V) is the most interesting phase for culinary purpose, but it is not in the thermodynamic equilibrium state. Therefore, cocoa butter has to be "forced" into this crystal phase that changes into an even more stable phase VI when stored for a long time (phase β (VI) is the most stable of all polymorphs of cocoa butter, with the highest density and highest melting temperature).

In the conventional tempering process, the melted cocoa butter is cooled down to 25°C – 27°C in order to form polymorphic nucloi of phase IV and V, which then grow into crystals; then the cocoa butter is heated again to 33°C and kept at this temperature for about 10 minutes, so that the crystals of the undesired phase IV melt, while the V crystals can form. Those crystals act as "seed crystals" that can grow slowly during the tempering time. These sufficient amounts of crystals are then further cooled down in order to remain in the desired phase V for the most part.

Is not it wonderful that chocolate makers discovered such a complex process without any knowledge of physical chemistry?

8.5.1.3 *Where the Beauty Lies*

Now, let us consider a recipe which includes ordinary butter, based on the "chocolate chantilly", that I introduced in 1995: here, the proposal is to make "butter chantilly".

In order to understand the process, let us start from the analysis of the transformations of milk (there will be no milk in the recipe, but it helps in the explaination): it is a dispersion of fat droplets in an aqueous solution. When milk sets, the fat "creams"; *i.e.,* because of their low density the fat droplets move upward, where they make a more concentrated emulsion called "cream". If you whip cream, the whisk introduces air bubbles, but if the system is cooled, the fat makes a solid network where the air bubbles are trapped. With this process, the initial emulsion is aerated before a solid networks forms, making the final whipped cream.

I proposed to make the same with other products than those from milk, and in particular water and chocolate: if you put 200 g water in a pan, with 225 g chocolate, heating it generates a chocolate emulsion, but if you put now the pan on ice cubes or in cold water and if you whip, you will get a chocolate mousse without eggs, the consistency being the same as for whipped cream (if the proportions of water and fat are the same as in cream).

To finish this section, you can do the same with butter: imagine putting in a pan an amount (about 200 g) of orange juice (aqueous solution) and butter (try also 280 g); heating will generate the emulsion, and whipping while cooling will make the "butter chantilly à l'orange".

8.5.2 The Question

Let us come back to the question of this section: is butter an emulsion? Many students do not understand why this question is given because they were often taught that mayonnaise is a prototype of a O/W emulsion (oil dispersed in water), and that butter would be W/O emulsion.

For mayonnaise, we could see that indeed, oil droplets are dispersed in an aqueous solution, but for butter? If so, why would the consistency of butter change with the temperature ? Let us focus on the words of the question, using the composition of butter. In particular, let us remember that, according to the IUPAC Gold Book, an emulsion is a dispersion of a liquid in another one. What are the "liquids" in butter?

8.5.3 Analysis of the Question

8.5.3.1 The Data is Introduced

As said, butter is made mostly of triglycerides, and water is allowed in it to the limit of 16% (w/w): this is from the FAO (2018), but let us remember that we could also have consulted the Codex Alimentarius, or "Food Code", *i.e.*, a collection of standards, guidelines and codes of practice adopted by the Codex Alimentarius Commission. The Commission, also known as CAC, is the central part of the Joint FAO/WHO Food Standards Programme and was established by FAO and WHO to protect consumer health and promote fair practices in food trade. It held its first meeting in 1963 (Codex Alimentarius, 2020). In the liquid aqueous phase, proteins and lactose are found, as well as mineral salts.

Now, about triglycerides, we could see that they all have different melting points, and this is why the solid content of butter decreases with temperature, as shown in Figure 8.14. But now we focus on the fact that the melting begins at about $-10°C$ and finishes at $55°C$ (Lopez and Ollivon, 2009).

8.5.3.2 Qualitative Model

What Figure 8.14 teaches is that the butter fat can be solid and liquid, in proportions that depend on the temperature. In particular, at temperatures below $-10°C$, the water component of butter would be solid (ice), and the fat content would be also entirely solid: this does not fit the IUPAC definition of emulsions: "A fluid colloidal system in which liquid droplets and/or liquid crystals are dispersed in a liquid. The droplets often exceed the usual limits for colloids in size." (IUPAC, 1972).

At temperatures higher than $60°C$, butter would not be an emulsion either because it would be a biphasic system (clarified butter).

And at room temperature? Part of the fat is liquid, and part of it is solid. How can these various parts be organized?

8.5.3.3 Quantitative Model

Remember that we want to make a calculation. Let us use the curve given in Figure 8.14 to see how much solid and liquid fat is present at "room temperature" (for example, $20°C$). Of course, there is a possibility of using a graphic system but this would be useless, as we need only an order of magnitude. Simply let us draw a line between $(-10°C, 100)$ and $(55°C, 0)$ and make an interpolation.

8.5.4 Solving

Sorry, but here, the calculation will be very simple. This will leave us some freedom for introducing a formalism that allows investigating the real physical structure of butter.

But first about the interpolation. Certainly, you can draw a line on a piece of paper on which you draw a diagram, but why not do it with a software? Considering the proportion p (in %) of solid as a function of temperature θ (in°C), let us consider the equation of the line between the points $(-10; 100)$ and $(0; 55)$ of the $(\theta°C; p)$ plane. The equation that we are looking for is:

$$p = a\theta + b$$

where a and b are two coefficients that we want to determine. To this end, we shall use the conditions:

$$eq1 := 100 = a \cdot (-10) + b: \quad eq2 := 0 = a \cdot (55) + b:$$

Solving it is easy, because the solve command can solve not only equations, but also systems of equations:

$$solve(\{eq1, eq2\}, \{a, b\}) \left\{ a = -\frac{20}{13}, b = \frac{1100}{13} \right\}$$

Now, we can write the equation:

$$p = -1.54\theta + 84.6$$

We simply put $\theta = 20°$ C in it to find the proportion of solids at this temperature:

$$-1.54 \cdot 20 + 84.6 = 53.80$$

You see: about 50%. It is not negligible, and we need to know where this solid is distributed!

In order to understand better the physical structure of butter, let us turn now to the formalism called DSF (dispersed system formalism), using symbols for phases, dimensions, and topological relationships. We have already seen that:

- for phases, we distinguish solids (S), oil (liquid fat, O), water (aqueous solutions, W) and gases (G);

- we use operators: "/" for a random dispersion, "+" when there is a coexistence of phases in another, "@" for inclusions, "σ" for superposition and "x" when two "continuous" phases of the same dimension are interleaved.

But we have not yet seen that we can describe the "dimensions" of the structures with the symbols D_0 for "dots" (when the three dimensions of an object are more than one order of magnitude less than a reference size), D_1 for "lines" (when two dimensions of an object are more than one order of magnitude less than a reference size), D_2 for "surfaces" (when one dimension of an object is more than one order of magnitude less than a reference size), and D_3 for objects whose three dimensions are of the order of magnitude of the reference size.

Using this formalism, we now have to apply to butter, in which there is water (W), liquid fat (oil, O) and solid fat (S). At room temperature, solid fat can be dispersed as $D_0(S)$ structures, or it can make a continuous network $D_3(S)$ (it is unlikely that it would make lines or sheets, because there is no preferred direction during the process).

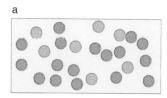

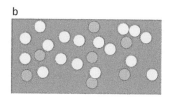

FIGURE 8.15 *Whatever the real microstructure of butter, it is either an emulsion/suspension (a) or a gel (b), but not an emulsion. Here, the solid fat is in orange, the liquid fat, or "oil", in yellow, and water is in blue.*

Let us distinguish the two cases:

- if the solid part of butter is under the form $D_0(S)$, then it has to be dispersed in a liquid phase, that can be water W or oil O; because the water proportion is low, it is more likely that we have $(D_0(S)+D_0(W))/D_3(O)$, *i.e.*, a double dispersion of water droplets and of solid fat in liquid fat (Figure 8.15a), and the system should be called an suspended emulsion, rather than an emulsion;

- if the solid part of butter is under the form $D_3(S)$, then because a liquid is trapped inside this solid network, the system is formally a gel (Figure 8.15b).

8.5.5   Conclusions and Perspectives

Finally, butter is not an "emulsion"; it is more a complex system, for which a DSF formula is needed.

8.6   WHEN WE LOOK AT MAYONNAISE SAUCE UNDER THE MICROSCOPE CAN WE SEE THE BROWNIAN MOTION?

8.6.1   An Experiment to Understand the Question

8.6.1.1   *First, the General Idea*

Certainly, when studying this question it is helpful to use a microscope and also to make a mayonnaise. The recipe of the sauce was already given many times previously: simply mix an egg yolk with a spoon of vinegar, then add oil slowly, while whisking.

When the sauce is made (smooth, high viscosity), take a tiny quantity of it, on the tip of a spatula, and spread a thin layer of it on a glass plate, with no cover slips (this could change the appearance of the soft emulsion). Then, put the plate on

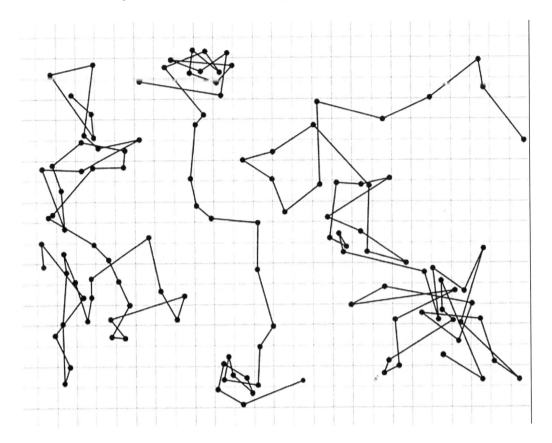

FIGURE 8.16 *An object immersed in water moves in a jittery motion. The picture here is a classic one, of an article in which the French physical chemist Jean Perrin was validating the interpretation of Brownian motion proposed by Albert Einstein (Perrin, 1913).*

the stage of the microscope, use the lowest magnification, and focus: do you see any random motion?

Use now a higher magnification, and observe: do you see erratic motion?

8.6.1.2 *Questions of Methods*

Before proposing an improved way of observing tiny displacements, let us observe that the "Brownian motion" was named after the English botanist Robert Brown (1773–1858), who discovered it in 1827: observing pollen grain with a microscope, he could see that organelles had a jittery motion. Later, he observed the same motion for inorganic particles and concluded that this had nothing to do with life. But Brownian motion became more popular after the beginning of the 20th century, when Albert Einstein established the mathematical equations for the movements of particles on the basis of the kinetic molecular theory (1905), with a definitive confirmation by the French physicist Jean Perrin (Figure 8.16).

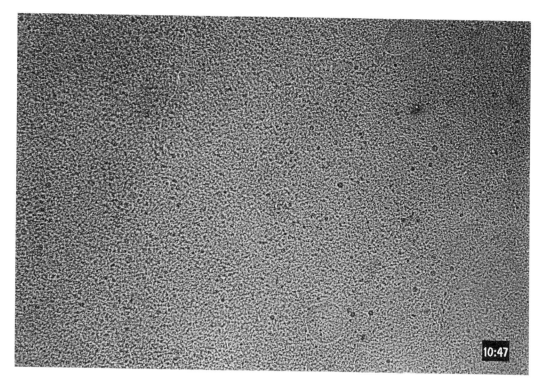

FIGURE 8.17 *Egg yolk under the microscope. Granules have a size between 0.3 and 2 μm.*

In order to observe Brownian motion, it is better to have suspended tiny particles in water, and a high magnification, such as 400 X.

For the particles, you can use those of milk: dilute two drops of milk in 5 mL of water, place one drop of the dilute solution on the slide with a cover slip on to of it. Use a bright field and a dark field.

If your microscope is not powerful enough, you can use your mobile phone to enlarge the picture to a greater level than that given by the ocular lens: fix the phone so that it does not move, and use the stronger zoom level of your camera (Deruyter and Brubacher, 2008).

8.6.1.3   *Where the Beauty Lies*

With your microscopic system, you are not restricted to milk. For example, you could begin with a drop of egg yolk and you would observe that it is made of "granules" dispersed in a plasma (Michalski *et al.*, 2004; Laca *et al.*, 2014), *i.e.*, a viscous aqueous solution (Figure 8.17).

You can also investigate the making of custard, for which milk and sugar are added to egg yolk. When you cook the whole mixture, proteins aggregate, as shown in Figure 8.18. And when "curdles" appear, if you overcook, the microscope can show that the microscopic aggregates are making denser groups that appear macroscopically.

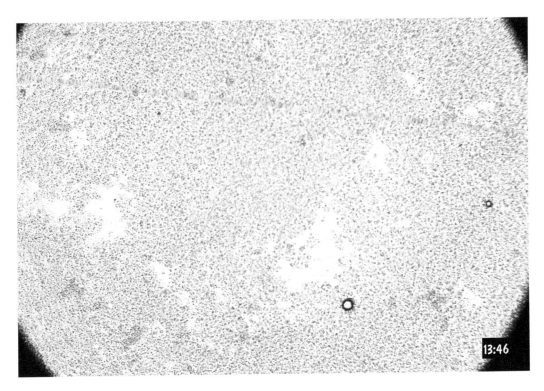

FIGURE 8.18 *A successful custard owes its viscosity to the making of aggregates from proteins.*

However, there is a folklore saying that you can "recover" a failed custard by shaking it with milk; certainly you get a result that resembles—macroscopically and microscopically (Figure 8.19)—a successful custard... but this shaking process does not suppress the flavor of cooked egg.

8.6.2 The Question

In Section 6.7, we discussed the conditions for using the Maxwell-Boltzmann distribution, and this one is a continuation of it. Some students know the Brownian motion and some do not, but it is a very important effect, as shown in Section 8.4, and today's science and technology can hardly escape meeting it sooner or later.

Because it involves randomness, some find it difficult to move from qualitative ideas toward calculation. But there are so many different ways of exploring it! Here we start from a now familiar sauce, but we go deeper into the use of microscopy.

8.6.3 Analysis of the Question

8.6.3.1 The Data is Introduced

There is no need to give numerical data, but it would be probably helpful to provide microscopes and ingredients in order to make the mayonnaise sauce, *i.e.*, one egg,

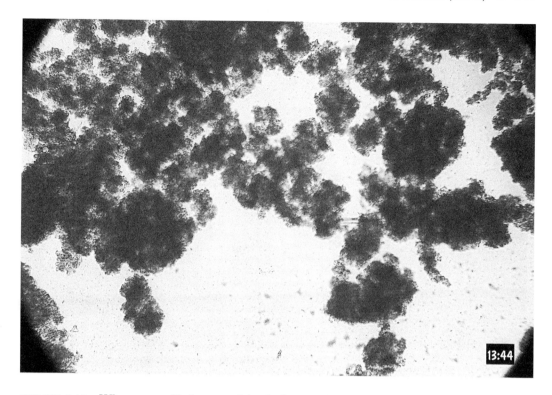

FIGURE 8.19 *When a curdled custard is shaken with milk, the macroscopic aggregates are dissociated.*

vinegar, oil, salt and pepper (no mustard, as it would transform mayonnaise into remoulade), a large bowl, and a whisk.

8.6.3.2 Qualitative Model

Let us now repeat the initial question: can we see a Brownian motion when observing a mayonnaise with a microscope? This motion is the result of a different number of collisions of an object with water molecules, on the opposite sides of the object.

In mayonnaise, there are many different kinds of objects, with radiuses ranging from 10^{-10} m (water molecules) to 0.1 mm (the size of the largest oil droplets), including proteins in solution, micelles (aggregates made of phospholipids), the granules from egg yolks (aggregates made of about 60% proteins and 40% lipids) (Anton and Gandemer, 1997), small oil droplets, and air bubbles.

But the question can be more quantitative: would we see the Brownian motion of oil droplets, for example? Or, expressed more precisely, below which size can we see this Brownian motion, that is due to the random motion of water molecules?

8.6.3.3 Quantitative Model

Let us first have a look at the system that we consider: an object "bigger" than water molecules, in the middle of a water liquid phase (Figure 8.20).

FIGURE 8.20 *In water, an object is moved in random directions by collisions with water molecules.*

The question of the "size" is important in this discussion, because one has the (right) feeling that a "big" object will be less moved by water molecules than a "small" one. And here is the key: because of the reflex that we discussed previously, of turning adjectives and adverbs into the answer to the question "how much?", we find a solution: "big" and "small" being adjectives, we have immediately to turn them quantitative. Here, the reference is the size of water molecules.

For our model, we need to introduce the mass M of an object, and its diameter d.

But we also know that the velocity of water molecules is important. Let us have:

- T for absolute temperature.

- v the average velocity of water molecules.

8.6.4 Solving

8.6.4.1 Looking for a Solving Strategy

The question is one of motion, and so kinetic energy is to be used. Also, one should know that Brownian motion occurs because at any time more molecules collide with the oil droplets on one side than on the other. Here, we want to estimate orders of magnitude of the kinetic of a particle suspended in water compared to the kinetic energy of water molecules.

8.6.4.2 Implementing the Strategy

Before calculating, let us observe that the interpretation of the Brownian motion was not simple, and the Einstein proposal came after much debate about the possibility of applying the kinetic theory of gases: as late at 1911, famous scientists did not accept the existence of atoms and molecules in the modern meaning that we understand today.

Einstein had three main ideas. The first was to consider that the gas equations could be extended to solutions, providing that osmotic pressure is considered instead of mechanical pressure; this allowed the application of the results of the kinetic theory of gases to molecules in solution, and as no restriction was placed on the size of the molecules, small suspended particles could be treated as huge molecules. The second idea was to use the Stokes drag force. And the third idea was to deal with displacement effect in a given time, independently of the path followed (Maiocchi, 1990; Gora et al., 2006; Genthon, 2021).

Here, we cannot go into such detailed description, but we want to have order of magnitudes of the various phenomena.

Let us consider an object of mass M being colliding with water molecules (mass m). In order to get an order of magnitude of the velocity that the object can get from water, we first calculate how many water molecules are on one side of the object.

Assuming that this object has the shape of a cube of side c and density ϱ, we could write:

$$\varrho = \frac{M}{c^3}$$

So that the side would be:

$$c = \left(\frac{M}{\varrho}\right)^{\frac{1}{3}}$$

If A is the area on one side:

$$A = c^2 = \left(\frac{M}{\varrho}\right)^{\frac{2}{3}}$$

Let us assume that water molecules would distribute regularly, with a square lattice, near the side of the cube, the distance between two molecules being the average distance d between water molecules in liquid water (see Annex 2 of Section 5.2 for the calculation of this). Then, the number of water molecules near one side of the cube would be:

$$N = \frac{A}{d^2} = \frac{\left(\frac{M}{\varrho}\right)^{\frac{2}{3}}}{d^2}$$

Let us now imagine that there is a difference of collisions between water molecules and the cube: this could be a difference by one molecule, or two, or three, up to N.

For only one water molecule at the temperature T, the kinetic energy is of the order of $\frac{3}{2}k_B T$, where k_B is the Boltzmann constant.

If this energy were entirely given to the cube, its kinetic energy E_c would become:

$$E_c = \frac{Mv^2}{2} = \frac{3}{2}k_B T$$

Knowing this kinetic energy, we can calculate the velocity of the object of mass M after the collision with one water molecule:

$$v = \sqrt{\frac{3\,k_R\,T}{M}}$$

8.6.5 Expressing the Results

8.6.5.1 The Formal Result Found is Written Down Again

For the number of water molecules on a side of the object:

$$N = \frac{c^2}{d^2}$$

with c the side of the object, and d the distance between water molecules.

For the velocity, instead of repeating simply what we found, we go on introducing the side of the object:

$$v = \sqrt{\frac{3k_BT}{M}} = \sqrt{\frac{3k_BT}{\varrho c^3}}$$

8.6.5.2 Finding Digital Data

First, in order to get a clearer idea of the number of collisions by water molecules, let us remember that we found an average distance of about 3×10^{-10} m between water molecules.

For the numerical calculations of the velocity, we consider first various masses, corresponding to particles of density equal to the one of water, and different sides (remember, we modeled cubes), from 10^{-3} to 10^{-9} m.

8.6.5.3 Introduction of Data in the Formal Solution

For the number of water molecules and for the velocity, we give the results in Table 8.2.

First, we see that the number of collisions can be huge, so that our assumption of disequilibrium by only one collision is very weak. Also our calculation shows that the

Table 8.2 The velocities of various objects through Brownian motion... if the disequilibrium would be only by one water molecule.

Side (m)	Number of water molecules near a side	Velocity (m/s)	Type of objects
10^{-3}	10^{13}	1.1×10^{-7}	large oil droplet
10^{-4}	10^{11}	3.5×10^{-6}	small oil droplets
10^{-5}	10^{9}	1.1×10^{-4}	very small oil droplets
10^{-6}	10^{7}	3.5×10^{-3}	granules
10^{-7}	10^{5}	1.1×10^{-1}	granules

velocities become observable only with very small objects, but we have also remember the assumptions that we used to make our calculations.

8.6.6 Conclusions and Perspectives

Of course, our calculation was simplistic, but remember that we simply wanted orders of magnitude. This should help us to interpret "stories", such as about the discovery of Brownian motion: it has been frequently said that Brown discovered a motion of pollen, but can you use data on pollen to guess if it is possible?

Indeed such stories are not exact: first, Robert Brown (1773–1858) was Scottish, rather than English; in 1828 he published the article *A brief account of microscopic observations made in the months of June, July and August, 1827, on the particles contained in the pollen of plants; and on the general existence of active molecules in organic and inorganic bodies* (Brown, 1828). In the *Edinburgh new Philosophical Journal*, he explained that he observed pollen seeds of the species *Clarkia pulchella* in water, and that he could observe, in these pollen particles, smaller particles, or granules, with a diameter of about 5 μm. These granules were moving. Indeed, this is more accurate: the granules in the pollen particles move, but the pollen particles themselves do not move much.

Now, if Einstein and Jean Perrin were clearly important in the history of physics, the right theory of Brownian motion was proposed in 1900 by the French mathematician Louis Jean-Baptiste Alphonse Bachelier (1870–1946), and only later by Albert Einstein and Marian Smoluchowski (Haw, 2002). Einstein theory states that the diffusion constant D for the Brownian motion of a particle of radius r, in a medium of viscosity η, is given by:

$$D = \frac{k_B \cdot T}{6\pi\eta r}$$

And the square of the fluctuation in position is equal to $D.t$.

Finally, it has to be observed that since solvent molecules are generally much smaller than colloidal particles, solvent molecules are regarded as a continuum and the influence of the motion of solvent molecules is combined into the equations of motion as a stochastic term (this is the "Langevin equation").

REFERENCES

Anton M, Gandemer G. 1997. *Composition, solubility and emulsifying properties of granules and plasm of egg yolk,* Journal of Food Science, 62 (3), 484–487.

Biedermann F, Nau WM, Schneider HJ. 2014. *The Hydrophobic Effect Revisited—Studies with Supramolecular Complexes Imply High-Energy Water as a Noncovalent Driving Force.* Angewandte Chemie International Edition, 53, 11158–11171.

Bierce A. 1911. *Devil's Dictionary,* https://gutenberg.org/ebooks/972, last access 2022-01-10.

Brown R. 1828. *A brief account of microscopical observations made in the months of June, July and August, 1827, on the particles contained in the pollen of plants; and on the general existence of active molecules in organic and inorganic bodies*, The Philosophical Magazine, 4, 161-173. https://doi.org/10.1080/14786442808674769

Carême MA. 1828. *Le cuisinier parisien*, Chez l'auteur, Paris, France.

Codex Alimentarius. 2020. www.fao.org/fi.e.ao-who-codexalimentarius/home/en/, last access 2022-01-10.

Conti Nabali V, Pezzotti S, Sebastiani F, Galimberti DR, Schwaab G, Heiden M, Gaigeot MP. 2020. *Wrapping up hydrophobic hydration: locality matters*, The Journal of Physical Chemistry Letters, 11, 4809–4816.

Cottrell TL. 1958. *The strengths of chemical bonds*, Butterworths, London.

De Gennes PG, Brochard-Wyart F, Quéré D. 2004. *Capillarity and wetting phenomena: drops, bubbles, pearls, waves*. Springer, Heidelberg, Germany.

Deruyter HH, Brubacher LJ. 2008. *High school students calculate Avogadro's constant using video projection of Brownian motion in milk*, Chem 13 News, 1–6, https://uwaterloo.ca/chem13news/sites/ca.chem13news/files/uploads/files/Page3_7from%20Jan_2008_Number%20353.pdf.

FAO. 2018. *Standard for butter CXS 279-1971*, Formerly Codex STAN A-1-1971. Adopted in 1971. Revised in 1999. Amended in 2003, 2006, 2010, 2018.

Genthon A. 2021. *The concept of velocity in the history of Brownian motion*, The European Physical Journal, https://doi.org/10.1140/epjh/e2020-10009-8, last access 2022-01-14.

Gora PF, Smoluchowski M, William R. 2006. *The theory of Brownian motion: a hundred year's anniversary*, Marian Smoluchowski Symposium, http://www.smoluchowski.if.uj.edu.pl/smoluchowski-2017

Grillo I. 2003. *Small-angle neutron scattering study of a world-wide known emulsion: Le Pastis*, Colloids and Surfaces A, Physicochemical and Engineering Aspects, 225(1–3),153–160.

Haw MD. 2002. *Colloidal suspensions, Brownian motion, molecular reality: a short history*, Journal of Physics: Condensed Matter, 14, 7769–7779.

IUPAC. 1972. *Manual of Symbols and Terminology for Physicochemical Quantities and Units, Appendix II: Definitions, Terminology and Symbols in Colloid and Surface Chemistry*, Pure and Applied Chemistry, 31, 577–606.

IUPAC. 2019a. *Emulsions*, https://doi.org/10.1351/goldbook, last access 2022-01-10.

Laca A, Paredes B, Rendueles M, Diaz M. 2014. *Egg yolk granules: separation, characteristics and applications in food industry*, LWT - Food Science and Technology, 59, 1–5.

Lavoisier AL. 1783. *Mémoire sur le degré de force que doit avoir le bouillon, sur sa pesanteur spécifique et sur la quantité de matière gélatineuse solide qu'il contient.* In *Lavoisier, œuvres complètes.* Expérience de novembre 1783, 33, 563–575.
This article demonstrate the smartness of Lavoisier. And it is about cooking!

Lopez C, Bourgaux C, Lesieur P, Riaublanc A, Ollivon M. 2006. *Milk fat and primary fractions obtained by dry fractionation. 1. Chemical composition and crystallisation properties*, Chemistry and Physics of Lipids, 144, 17–33.

Lopez C, Briard-Bion V, Bourgaux C Pérez J. 2013. *Solid triacylglycerols within human fat globules: β crystals with a melting point above in-body temperature of infants, formed upon storage of breast milk at low temperature*, Food Research International, 54, 1541–1552.

Lopez C, Ollivon M. 2009. *Triglycerides obtained by dry fractionation of milk fat2. Thermal properties and polymorphic evolutions on heating*, Chemistry and Physics of Lipids, 159, 1–12.

Maiocchi R. 1990. *The case of Brownian motion*, BJHS, 23, 257–283.

Marin F. 1758. *Les dons de Comus*, Pissot, Paris, France.

Michalski MC, Camier B, Briard V, Leconte N, Gassi JY, Goudédranche H, Michel F, Fauquant J. 2004. *The size of native milk fat globules affects physico-chemical and functional properties of Emmental cheese.* Le Lait, INRA Editions, 2004, 84 (4), 343–358.

NCBI. 2002. *Problems, in biochemistry*, 5th edition, W.H. Fremann, New York, USA. https://www.ncbi.nlm.nih.gov/books/NBK22504/.

Perrin J. 1926. *The Nobel prize in physics 1926*, https://www.nobelprize.org/prizes/physics/1926/summary/, last access 2022-01-10.

Pierotti RA. 1976. A scaled particle theory of aqueous and nonaqueous solutions. Chemical Reviews, 76, 171–726.

Reif F. 1967. Statistical physics, Berkeley Physics Course, McGraw-Hill, New York, 5.

Silverstein TP. 1998. *The real reason why oil and water don't mix*, Journal of Chemical Education, 75(1), 116–118.

Taillevent. 2000. *Viandier*, Justus-Liebig Universität Giessen. https://www.uni-giessen.de/fbz/fb05/germanistik/absprache/sprachverwendung/gloning/tx/vivat.htm, last access 2022-01-10.

This H. 2003. *L'effet Pastis*, https://pierregagnaire.com/pierre_gagnaire/travaux_detail/66, last access 2022-01-10.

This H 2009. *Molecular Gastronomy, a chemical look to cooking.* Accounts of Chemical Research, 42(5), 575–583.

This H. 1996. *Can a cooked egg white be uncooked ?*, The Chemical Intelligencer (Springer Verlag), (10), 51.

This H. 2010. *Les précisions culinaires*, Quae/Belin, Paris, France.

This H. 2016a. *Methodological advances in scientific publication*, Notes Académiques de l'Académie d'agriculture de France, 8, 1–26.

This H. 2016b. *Statgels and dynagels*, Notes Académiques de l'Académie d'agriculture de France, 12, 1–12.

This vo Kientza H. 2021. *Sauces*, Handbook of molecular gastronomy, CRC Press, Boca Raton, 495–498.

This H, Méric R, Cazor A. 2006. *Lavoisier and meat stock.* Comptes rendus de l'Académie des sciences, Chimie, 11–12, 1511–1515.

CHAPTER 9

Gels

I N THE TITLE of this chapter, you do not see any DSF formula, after the word "Gels", and the reason for it is that there are too many different kind of gels, even if one considers the simplest ones, with a liquid phase in a solid (IUPAC, 2019): "Non-fluid colloidal network or polymer network that is expanded throughout its whole volume by a fluid."

This definition goes along with "Notes":

"A gel has a finite, usually rather small, yield stress.

A gel can contain:

1. a covalent polymer network, *e.g.*, a network formed by crosslinking polymer chains or by non-linear polymerization;

2. a polymer network formed through the physical aggregation of polymer chains, caused by hydrogen bonds, crystallization, helix formation, complexation, etc., that results in regions of local order acting as the network junction points. The resulting swollen network may be termed a thermoreversible gel if the regions of local order are thermally reversible;

3. a polymer network formed through glassy junction points, *e.g.*, one based on block copolymers. If the junction points are thermally reversible glassy domains, the resulting swollen network may also be termed a thermoreversible gel;

4. lamellar structures including mesophases, *e.g.*, soap gels, phospholipids and clays;

5. particulate disordered structures, *e.g.*, a flocculent precipitate usually consisting of particles with large geometrical anisotropy, such as in V_2O_5 gels and globular or fibrillar protein gels."

Finally, using the DSF can help you in exploring gels and their properties, but this would be another story (This, 2016; This vo Kientza, 2021).

DOI: 10.1201/9781003298151-9

9.1   HOW MUCH GEL CAN BE MADE WITH A KNOWN MASS OF GELATINE?

9.1.1   An Experiment to Understand the Question

9.1.1.1   First, the General Idea

Here, in this chapter about gels, let us investigate the systems that are given this name:

1. cook a piece of skate in water (simply enough to cover the fish), under a lid, until boiling; when the skate is cooked, put the pan in the fridge and after one night, observe that the transparent colorless liquid turned into a transparent solid (the material does not flow);

2. do the same, but with a foot of veal or pork, with a processing time of five hours, in a pan with a lid, water only simmering; and after cooling in the fridge, observe the same "gelification";

3. put an egg white in a bowl, and heat at low/medium heat, in a microwave oven: you observe the solidification of the egg white, but of course at a rate much faster than in the two previous cases; initially transparent (and yellow greenish), it is transformed into a (white) material that does not flow;

4. repeat the same as 3, but before heating the egg white, dilute it with twice the volume of water; observe that the gel that forms is softer than the ordinary cooked egg white;

5. in a bowl, put 100 g of cold water, and soak 20 g of gelatine for about 3 min; then bring to the boil, and store in the fridge: observe after some hours that the transparent liquid turned into a soft solid;

6. repeat the same as 5, but with only 10 g of gelatine for 100 g of water: observe the gel is softer;

7. repeat the same as 5, but with only 5 g of gelatine for 100 g of water: observe that the gel is even softer;

8. repeat the same as 5, but with only 1 g of gelatine for 100 g of water: observe that the liquid does not set.

You see: in the previous experiments (except for #6), there was always the transformation of a liquid into a soft solid, *i.e.*, a material that does not flow. Of course, such a solid is not like a piece of metal, and we remember that it contains a lot of water (in the cases considered above); nonetheless, there is a difference with a liquid, which would flow more or less easily.

In the past, the gelification was called a "coagulation", but today this last term is given a different meaning: "When a sol is colloidally unstable (*i.e.* the rate of aggregation is not negligible) the formation of aggregates is called coagulation or flocculation" (IUPAC, 1972). Here, a sol is defined as "A fluid colloidal system of two or more components, *e.g.*, a protein sol, a gold sol an emulsion, a surfactant solution above the critical micelle concentration".

9.1.1.2 Questions of Methods

Some courses in physical chemistry distinguish "chemical gels" and "physical gels": the chemical gels are irreversible (*e.g.*, the coagulated egg white), and the physical gels are thermo-reversible (*e.g.*, the gelatine gels). For example, Figure 9.1 shows the general description of a gel made from proteins such as in egg whites.

But how do the strands of the network bind? For discussing this question, Figure 9.2 (again!) is useful.

This continuous scale does not show the clear limit between reversible and non-reversible gels, and this is why, in 1995, I wanted to "uncook" an egg, using the following analysis (This, 1996):

1. If egg white coagulates (and makes gels), it is because of chemical bonds.

2. What is the most energetic one? As proteins can bind through van der Waals forces, hydrogen bonds, disulfide bridges, and though unlikely, through covalent bonds, then disulfide bridges or covalent bonds are the main possibilities.

3. Disulfide bridges are between sulfur atoms of the cysteine residues, and they are the result of an oxidation (Figure 9.3).

4. Accordingly, the addition of a reducing agent to cooked egg white (for example, sodium borohydride $NaBH_4$) would "uncook" the egg... and it does!

9.1.1.3 Where the Beauty Lies

Now that we understand the making of a gel better, let us make a culinary interesting "dirac", *i.e.*, a gellified emulsion. Below, the process indicates how to make first the emulsion, and then turn it into a gellified system, using the egg white proteins at the interface of oil droplets.

1. Place an egg white into a vessel.

2. Using a whisk, whip it until you get a foam.

3. Add oil while whipping, until you get a white creamy texture, such as the one of a sauce mayonnaise (with about 200 g oil).

4. Add 50 g of sugar, a food colorant, a pinch of salt, some lemon juice (to taste), and some vanilla flavor.

5. Distribute the emulsion in cups.

6. Cook in the microwave oven (full energy) until you see a 30% expansion.

7. Serve hot.

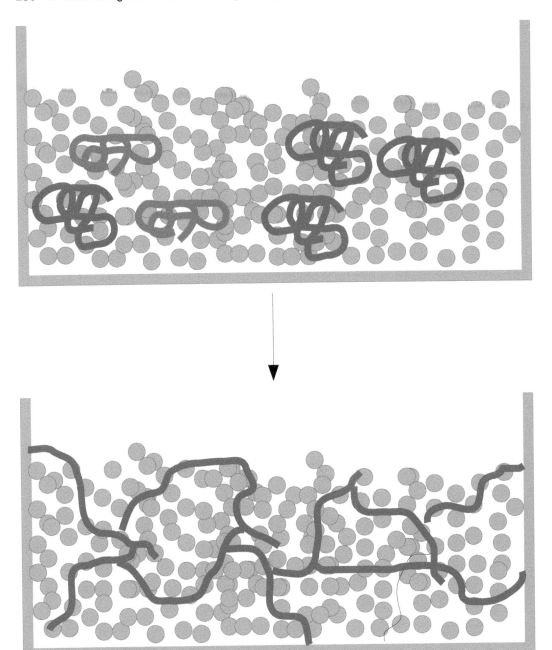

FIGURE 9.1 *The making of a gel from proteins that are denatured by heat. The proteins (red) make a continuous network that traps water molecules (blue dots).*

9.1.2 The Question

In the kitchen, gelatine is often used for making gels. It can be supplied as a powder, or sheets, and it is always of animal origin, in contrast to other gelling agents such as agar-agar or carrageenans, that are extracted from algae. Different types of gelatine

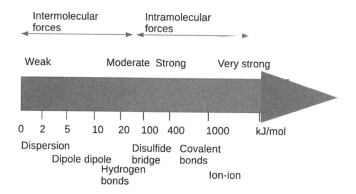

FIGURE 9.2 *The energy of the various bonds. It is helpful to remember that on a scale where the covalent carbon-carbon bond would be at 100, the van der Waals bonds would be at about 1 and the hydrogen bonds at 10.*

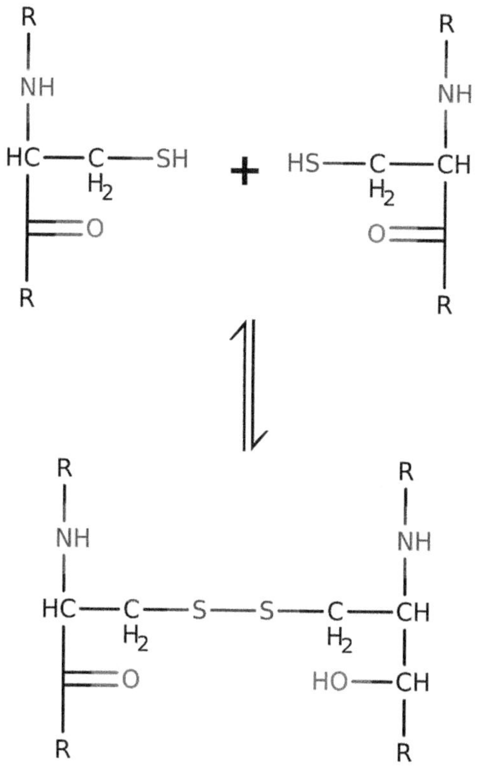

FIGURE 9.3 *The making of a disulfide bridge, from two thiol groups of neighboring proteins is an oxidation. You can break such a bond using a reducing agent, such as sodium borohydride (use about 1 teaspoon for 1 cooked egg white... and don't eat it!).*

exist, with different gelling properties (*e.g.*, temperature of gelation, quantity needed), but the order of magnitude is always the same:

- gellification temperature at about 36°C;

- 5% (mass) is needed for making a gel with water.

Can we calculate the amount of gel that we can make with a known mass of gelatine?

9.1.3 Analysis of the Question

9.1.3.1 *The Data is Introduced*

To the data given above, it can be added that gelatine is made from collagen, the main fibrous protein constituent in bones, cartilages and skins of animals. To date, up to 27 different types of collagen have been identified, but type I collagen is the most widely occurring collagen in connective tissue. Collagen molecules are composed of three chains intertwined in the so-called collagen triple-helix. This particular structure, mainly stabilized by intra- and inter-chain hydrogen bonding, is the product of an almost continuous repeating of the Gly-X-Y- sequence, where Gly stands for a residue of glycine, and X and Y are mostly proline and hydroxyproline residues respectively (Asghar and Henrickson, 1982). After extraction, gelatine is made of chains of different length, that depends on the extraction process. A molar mass of 150,000 g can be assumed for the calculations (Elharfaoui *et al.*, 2007).

9.1.3.2 *Quantitative Model*

Gelatine gels are bicontinuous systems, with water and a protein network. Either one knows that a minimum of 5% gelatine (mass) is needed or it is not known. We shall consider both cases.

1. Firstly, let us assume that we do not know how to use gelatine practically.

   Let m be the mass of gelatine used to make a gel. If M is its molar mass, the number of moles n of gelatine is:

   $$n = \frac{m}{M}$$

   And the number N of gelatine molecules is:

   $$N = n\, N_A = \frac{m}{M} N_A$$

   Let us assume that these molecules make a cubic network (lattice calculation): there are 12 edges per cube, but each edge is shared by 4 cubes; *i.e.*, only 3 edges belonging to each cube. So the number of cubes is obtained by dividing the number of gelatine molecules by 3:

   $$n_c = \frac{N}{3} = \frac{m N_A}{3M}$$

The total volume of these cubes is:

$$V = n_c v$$

where v is the volume of one cube.

This volume v is:

$$v = c^3$$

where c is the side of one cube, or the length of a protein (we first assume that they are fully extended, which is certainly not true).

Now we need to find the value of c. It calls for considering that proteins are sequences of residues of amino acids. Let us consider that a gelatine molecule is made of n_a amino acid residues, having each a length λ, i.e., a total length:

$$c = n_a \lambda$$

Thus:

$$V = \frac{mN_A}{3M}(n_a\lambda)^3$$

We calculate exceptionally the numerical result here. For example, we calculate the volume of gel that can be obtained with 10 g of gelatine. For λ, we assume 3 covalent bonds, and for n_a, we divide the molar mass by the average molar mass of an amino acid residue (about 120)

$$subs\left(m = 10e - 3, N_A = 6e23, M = 150000e - 3, \right.$$

$$\left. n_a = 120, \lambda = 3 \cdot 1.5e - 10, \frac{mN_A}{3M}(n_a\lambda)^3\right)$$

$$2.10 \times 10^{-3}$$

This value is in m^3: it is slightly more than 2 L.

2. If we know that one can make a gel from about 5% gelatine:

We now start from a proportion p of gelatine. For a mass m, a mass M of water with:

$$M = \frac{1-p}{p}m$$

$$subs(m = 10e - 3, p = 0.05, \frac{1-p}{p}m)$$

$$1.90 \times 10^{-1}$$

The mass is in kg, which corresponds to a volume much less that the one calculated in (1).

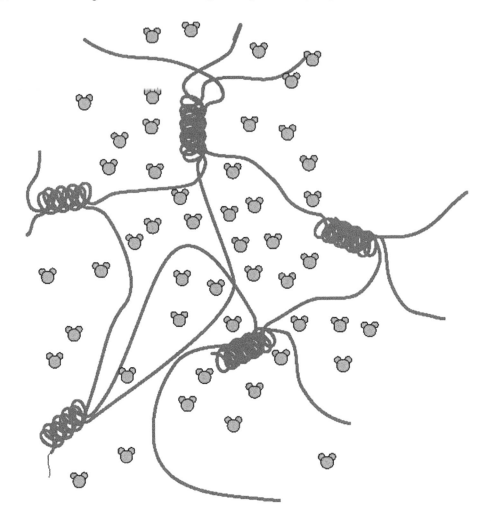

FIGURE 9.4 *Gelatine gels are created when the molecules form short segments of a triple helix. This process can be very long: after weeks, the gelatine gels go on evolving (Djabourov, 1991).*

9.1.4 Conclusions and Perspectives

With the first calculation, we found a volume of about 2 L, and with the second one, we found less, 0.2 L. The fact that our first calculation led to a larger volume of gel can be explained by the unlikely assumption that we made, of fully extended proteins, but also the model is not the right one: gelatine molecules do not make a cubic network, but instead bind by three, making segments of triple helices (Figure 9.4).

REFERENCES

Asghar A, Henrickson RL. 1982. *Chemical, biochemical, functional, and nutritional characteristics of collagen in food systems.* In Chischester CO, Mark EM, Stewart

GF (Eds.), Advances in food research, Academic Press, Cambridge, MA, 28, 232–372.

Djabourov M. 1991. *Gelation, a review.* Polymer International, 25, 135–143.

Elharfaoui N, Djabourov M, Babel W. 2007. *Molecular weight influence on gelatin gels: structure, enthalpy and rheology,* Macromolecular Symposia, 256, 149–157.

IUPAC. 1972. *Coagulation (flocculation),* PAC, 31, 609.

IUPAC. 2019. *Gel,* https://doi.org/10.1351/goldbook, last access 2022-01-10.

This H. 1996. *Can a cooked egg white be uncooked ?,* The Chemical Intelligencer (Springer Verlag), 10, 51.

If you want to see it, there is a video here, in three parts:

- cooking: https://www.youtube.com/watch?v=oXwderdkbqc
- analysis: https://www.youtube.com/watch?v=RjB_XoI2Ovs
- uncooking: https://www.youtube.com/watch?v=8ADrUUylLvw

This H. 2016. *Statgels and dynagels,* Notes Académiques de l'Académie d'agriculture de France, 12, 1 12.

If you want to learn more about gels, you can use this article: it shows that there are many more possibilities than you can imagine.

This vo Kientza H. 2021. *Gels,* Handbook of molecular gastronomy, CRC Press, Boca Raton, 375–380.

CHAPTER 10

Suspensions

Y OU KNOW THE GAME now: a new chemical word, a new consultation of the
IUPAC GoldBook, this time for "suspensions": "A liquid in which solid particles
are dispersed" (IUPAC, 2019).

However, things can be more complex, here as for gels, because there are different
liquids and different kind of solids, and different organizations of the dispersed solids.

10.1   HOW "THICK" CAN A VELOUTÉ BE?

10.1.1   An Experiment to Understand the Question

10.1.1.1   First, the General Idea

In classical French culinary art, there is a famous sauce called a velouté (Montagné,
1934). Recipes can be very complex, but basically veloutés are obtained by cooking
meat with onions and carrots, meat stock, and a "roux", *i.e.*, butter cooked with
flour. Of course, many other ingredients can be added such as milk, cream, egg yolks,
shallots, parsley, etc. And depending on the particular ingredients, you have different
sauces such as sauce Robert, sauce blanche, etc.

Here we propose to make a velouté in the following way:

1. in a pan, put four spoons of flour and the same volume of butter, and heat at
   very low heat: the mixture should blister softly;

2. add 250 g of milk;

3. heat gently while whipping (the goal is not to foam, but to disperse possible
   lumps;

4. when the sauce is homogeneous, add salt, pepper, nutmeg to taste, plus a dice
   of shallots, thyme, and cream.

You can serve it with turnips or potatoes boiled in salted water, for example, or
with fish.

DOI: 10.1201/9781003298151-10

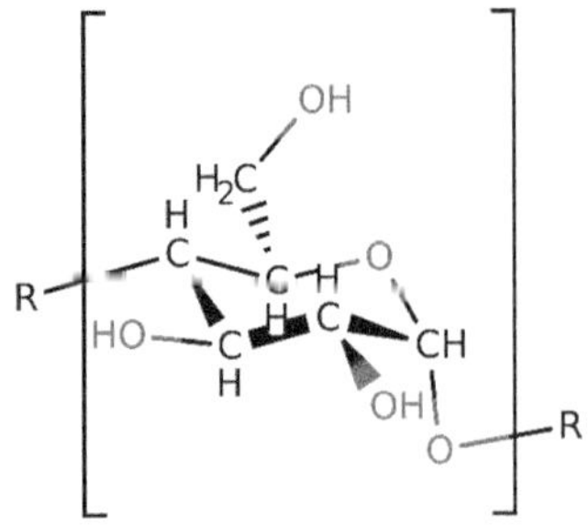

FIGURE 10.1 *Amylose molecules are polymers of D-glucose.*

10.1.1.2 *Questions of Methods*

The recipe given above deserves closer examination: why does it thicken after some time? Here again, a microscope can be useful, and you can begin by observing raw flour: you see starch granules. If you add water, the appearance does not change, but the observation is easier because the starch granules are dispersed in water ("suspension").

Now, if you heat the mixture, you will be able to observe the expansion of starch granules. However, the contrast in such a picture is weak—in particular when the granules have expanded a lot—and a colorant is helpful. The traditional one for starch is a solution of iodine (more precisely, you can use Lugol's iodine, with potassium iodide and iodine in water), because it is characteristic of one of its component: amylose.

What is it? Starch granules are organized in concentric layers made of two compounds: amylose and amylopectin, respectively linear and branched polymers of D-glucose (Figure 10.1). Iodine reacts with amylose that makes helices, including rows of six iodine atoms, and this turns the brown color to a characteristic blue (Figure 10.2) (Immel and Lichtenthaler, 2000).

Now, to finish here about methods, I invite you to make a model velouté from water and starch and to put it in the fridge. After one night of storage, you will observe that the sauce has gellified and that water droplets appear at the surface: upon cooling, starch molecules reassociate in a complex recrystallization process known as retrogradation, which is often associated with water separation from the gel (syneresis) (Tovar *et al.*, 2002).

10.1.1.3 *Where the Beauty Lies*

Sauces are strange systems, because they have intermediate viscosity between liquids and solids. Often, they make it easier for pieces (of meat, vegetables, etc.) to slip down the throat, but also their flavor is more readily accessible than when the same amount is within a food matrix: as soon as solid pieces covered with sauce are in the mouth, the odorant and taste compounds (among others) from the sauce can reach their receptors in the nose (after evaporation) or in the mouth (after dissolution in

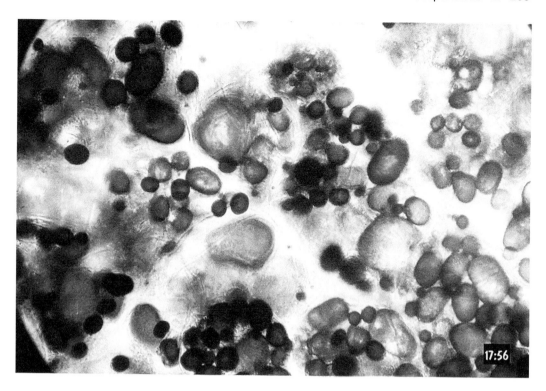

FIGURE 10.2 *Starch granules appear blue when they have been colored with brown iodine (here in a potato tissue). They are 30–100 micrometers in size.*

saliva); and it takes mastication to release the flavor compounds from the inside of the pieces, and provide other sensations.

The importance of the sauce in dishes, from the flavor point of view, is probably the reason why cooks give them a lot of flavor, using "reductions" (boiling animal or plant tissues for a long time) in order to concentrate the extracted compounds responsible for flavor), or adding various herbs and spices.

Now, an important culinary question about sauces is "thickening them"; *i.e.*, increasing their viscosity, also called liaison: often, cooks start with a liquid to which they want to impart more viscosity. To this end, they can emulsify liquid fat (as in mayonnaise), or expand starch (as in veloutés), or coagulate the proteins of egg yolk (custard) or blood (civets). There seems to be an almost infinite culinary diversity, but the difference between using chicken broth and wine is mainly a question of flavor, as the two liquids are aqueous solutions; and, as we could see in the previous chapter, gels from gelatine are the same kind of physical systems as gels made with agar-agar. With sauces, one can easily recognize that only a handful of physical chemistry principles apply. How many kinds of different physical chemistry systems do exist in sauces?

Using the so-called dispersed system formalism (DSF, see Section 8.5), it was shown that for French cuisine, the 450 sauces can be reduced down to 23 kinds, which can be organized by order of complexity (*e.g.*, a liquid is simpler than an emulsion) (This vo Kientza, 2021). But in the list, some simple systems are not present: the empirical work of cooks of the past did not lead them to discover some systems that could have made new sauces. In particular, there was no foamy velouté, even if such a system is easy to produce: in effect, you make a velouté, you make a foam (imagine whipping cream) and you mix the two (This, 2004). This new kind of sauce was introduced under the name "gay-lussac", from the French chemist Joseph Louis Gay-Lussac (1778–1850), famous for many works, but in particular for his determination of the relationship between pressure and temperature of gases.

10.1.2 The Question

As mentioned viscosity is an important characteristic of sauces. There is a lot of scientific literature about this, with many models. For example, because many sauces are non-Newtonian (the viscosity is changing with the flow), there are equations between the shear stress and the shear rate for these fluids.

Oups, you forgot about all that? Throughout this book, we have preferred to refer those who are missing information to their courses but we can make an exception here because this chapter is short. And because sauces are intermediate between solids and liquids, we shall consider both.

For solids, first, let us consider a cylinder of height H and section of area A, with the symmetry axis vertical. If we press the upper surface along the symmetry axis with a force of intensity F, the cylinder deforms: its height decrease (and if the material is incompressible, the section increases). In order to characterize this quantitatively, we use the "normal stress", a pressure (in pascals, Pa, or N.m^{-2}):

$$\sigma = \frac{F}{A}$$

But because a long cylinder will probably deform differently from a small one, for the same stress, we use the strain, defined as the ratio of the deformation ΔH by the initial height H_0:

$$\varepsilon = \frac{\Delta H}{H_0}$$

The simplest equation between stress and strain is of course when they are proportional, which is called Hooke equation, from the British physicist Robert Hooke (1635–1703):

$$E = \frac{\sigma}{\varepsilon}$$

The proportionality factor E is the Young modulus. And it is the equivalent, for a spring, to its stiffness. Of course, what we did is a simplistic approach, the uniaxial one, and physicists worked since the 17th century to improve this simple result (in particular A is changing), you can easily see that for a large deformation, the equation cannot hold, and more complex treatments have to be used.

For a shear stress, the picture is different (Figure 10.3). Now, for the simplest case, the shear stress σ is proportional to the shear strain:

$$\sigma = G\gamma$$

with:

$$\gamma = \frac{\Delta L}{H_0}$$

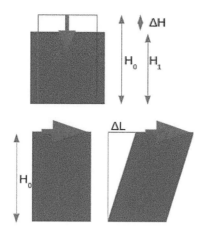

FIGURE 10.3 *The deformations of a solid.*

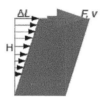

FIGURE 10.4 *The deformation of a liquid, when a shear force is applied at the surface (imagine pushing a board floating at the surface of liquid honey).*

For liquids, there is certainly a possibility to apply all this if they are in a closed vessel, but more generally, they flow, and the equations have to link the force applied to the flow. Imagine that you push a board floating at the surface of a viscous liquid, such as honey (Figure 10.4). Again, the shear stress can be defined as the ratio of the applied force by the area of the initial slab of liquid:

$$\sigma = \frac{F}{A}$$

And the deformation γ can be written:

$$\gamma = \frac{\Delta L}{H}$$

But we have to remember that the ΔL changes with time, and it is more appropriate to use the variation of γ with time. This is written frequently:

$$\dot{\gamma} = \frac{\mathrm{d}}{\mathrm{d}t}(\gamma)$$

And for the simplest case:

$$\dot{\gamma} = \frac{v}{H}$$

With all this, we can finally write the simplest equation (Newton's equation) between shear stress the $\dot{\gamma}$:

$$\sigma = \eta\dot{\gamma}$$

with the viscosity η.

This equation, as said, is for very simple liquids, but for sauces, it does not hold, and other expressions were found. For example, one important equation is the "power equation" also called Ostwald Waele model (Colin-Henrion *et al.*, 2007):

$$\sigma = K(\dot{\gamma})^n$$

Another one is the Herschel-Bulkley model:

$$\sigma = \sigma_0 + K(\dot{\gamma})^n$$

But there are many others, with the names of Casson, Carreau, Sisko, Mizrah-Berk...

Certainly, many sauces contain several constituents and are complex systems, with particles in suspension or large molecules in solution. Some of the rheology models are purely empirical, while others are based on the hydrodynamic approach of Einstein or the network approach (Einstein, 1906). All these correlations predict the viscosity of polymer solutions or suspensions of noninteracting species.

When the dispersion is of hard spheres, with no interaction (dilute regime), the Einstein relationship can be applied, keeping in mind that it is a first approach.

10.1.3   Analysis of the Question

10.1.3.1   The Data is Introduced

It is proposed to discuss the question for 15 g of flour added to 0.5 L of milk.

10.1.3.2   Quantitative Model

The viscosity η of a suspension, for the dilute regime, with hard spheres suspended in a simple Newtonian liquid, was found by Albert Einstein to be:

$$\eta = \eta_0(1 + 2.5 \cdot \phi)$$

where η_0 is the viscosity of the continuous phase, and ϕ is the volume fraction of suspended spheres.

But wait: the volume fraction? If you have a doubt about its definition, you can look at the I.U.P.A.C documents and you will find that it is the ratio of the volume of all dispersed solids by the total volume.

10.1.4   Solving

10.1.4.1   Implementing the Strategy

Here, we propose to simply introduce numerical data in the Einstein equation before having an important discussion about the limits of the formula.

10.1.4.2 Finding Digital Data

For the application of Einstein's equation, we need to know:

- the viscosity of the continuous liquid medium,

- the volume fraction of the dispersed phase.

For water at 20°C: $\eta_0 = 10^{-4} \text{Pa} \cdot \text{s}$
For milk: $\eta_0 = 19.74\,10^{-4} \text{Pa} \cdot \text{s}$ (Belitz and Grosch, 1997).
About the dispersed solids, let us consider flour. It is made of starch and other materials (such as proteins), but here, we shall assume that we are using only starch granules. Their diameter is between 2 and 35 μm (Olkku and Rha, 1978), and their density (for air-equilibrated starch of 10–15% moisture content) is 1.5 g/cm^3, *i.e.*, 1500 kg/m^3 (Dengate *et al.*, 1978).

10.1.4.3 Using the Data

We shall use the Einstein's equation for calculating the viscosity for four cases:

1. Non-expanded starch granules in water:

    For the liquid, we decided 15 g of solid material in 0.5 L (5×10^{-4} m^3) of liquid. For the solid, with 15 g, this makes 10^{-5} m^3. We find (units are Pa · s):

    $$subs\left(\eta_0 = 10^{-4}, \phi = \frac{1e-5}{5e-4}, \eta = \eta_0\left(1 + 2.5 \cdot \phi\right)\right)$$
    $$\eta = 1.05 \times 10^{-4}$$

2. Non-expanded starch granules in milk:

    Remember that milk is an emulsion, containing about 36 g of fat for 1 L (Anses, 2022). Is this enough to explain the viscosity of milk?

    If we assume a density of 900 kg · m^{-3}, we calculate a volume given by:

    $$v = \frac{m}{\varrho} = \frac{36e-3}{900}$$

    Thus:
    $$v = \frac{36e-3}{900} = v = 4.0 \times 10^{-5}$$

    And now, for the viscosity:

    $$subs(\eta_0 = 10^{-4}, \phi = \frac{4e-5}{1e-3}, \eta = \eta_0(1 + 2.5 \cdot \phi))$$
    $$\eta = 1.10 \times 10^{-4}$$

    This is much less than $19.74\,10^{-4} \text{Pa} \cdot \text{s}$. But remember that milk is not only fat droplets dispersed in water: there are also sugars dissolved in the aqueous phase, dispersed casein micelles... (Belitz and Grosch, 1997). Moreover, fat globules are not hard spheres.

And of course, if we calculate the viscosity using milk for a dispersion medium for dry particles, we can predict that the change (compared to milk) will not be large:

$$subs(\eta_0 = 19.7 \cdot 10^{-4}, \phi = \frac{1e - 5}{5e - 4}, \eta = \eta_0 (1 + 2.5 \cdot \phi))$$

$$\eta = 2.07 \times 10^{-3} \tag{10.1}$$

3. Swollen starch granules in water:

Swollen starch granules are much bigger than dry ones, but they are not hard spheres. As for milk, the result that we calculate is at best an order of magnitude. Here, we need some preparation before applying the formula, because the expansion of granules goes on with a change in density.

Let us first estimate the extent of their expansion. Figure 10.2, for example, shows that the diameter can increase about 10 times in cooked potatoes (the swollen granules fill completely the cells). And then the volume is increased as the cube of 10, *i.e.*, 1000.

During this expansion, water is absorbed, so that the density is:

$$\varrho_{\exp} = 0.001\varrho_{dry} + 0999\varrho_{water}$$

Numerically, we find:

$$\varrho_{\exp} = 0.001 \cdot 1.5 \cdot 10^3 + 0.999 \cdot 1e3$$
$$\varrho_{\exp} = 1000.50$$

No surprise: this is almost the same density as water.

Using this value, we can now use the Einstein formula (again, with wrong assumptions):

$$subs(\eta_0 = 10^{-4}, \phi = \frac{1e3 \cdot 1e - 5}{5e - 4}, \eta = \eta_0 (1 + 2.5 \cdot \phi))$$

$$\eta = 5.10 \times 10^{-3}$$

Here, fortunately, we see an increase of viscosity: remember that expanded granules are very large.

4. Swollen starch granules in milk (same expansion):

$$subs(\eta_0 = 19.7 \cdot 10^{-4}, \phi = \frac{1e3 \cdot 1e - 5}{5e - 4}, \eta = \eta_0 (1 + 2.5 \cdot \phi))$$

$$\eta = 1.0 \times 10^{-1}$$

With swollen starch granules, the calculation shows an obvious increase in viscosity, but do you think that the 10-fold increase in diameter is realistic?

10.1.5 Conclusions and Perspectives

A good practice, when using such a crude approximation for calculation, is to go online and look for experimental values recorded for a similar system, in similar conditions.

Regardless, chapter is an invitation to test the Einstein's equation with very different systems: custard (there are as well granules and oil droplets dispersed in water), purée, etc. Can you try to find out when the formula is no longer valid?

REFERENCES

Belitz HD, Grosch W. 1987. *Food chemistry*, Springer Verlag Berlin, Heidelberg, p. 513.

Anses. 2022. Milk, whole, UHT, Ciqual, https://ciqual.anses.fr/#/aliments/19023/milk-whole-uht, last access 2022-01-26.

Colin-Henrion M, Cuvelier G, Renard CMGC. 2007. *Texture of pureed fruit and vegetable foods*, Stewart Postharvest Review, 5(3), 1–14.

Dengate HN, Baruch DW, Meredith P. 1978, *The density of wheat starch granules: a tracer dilution procedure for determining the density of an immiscible dispersed phase*, Starch/Starke, 30(3), S 80–84.

Einstein A. 1906. *Eine neue Bestimmung der Molekul-dimensionen*. Annalen der Physik, 19, 289–306.

Immel S, Lichtenthaler FW. 2000. *The hydrophobic topographies of amylose and its blue iodine complex*, Starch/Stärke, 52(1), 1–8.

IUPAC. 2019. *Suspension*, https://doi.org/10.1351/goldbook, last access 2022-01-10.

Montagné P. 1934. *Larousse gastronomique*, Larousse, Paris.

Such books can be trusted for classic recipes, but certainly not for "explanations" of phenomena that you can observe during culinary preparations. In particular, this book transmits many tips, proverbs, sayings, old wive tales that have been refuted by molecular gastronomy.

Olkku J, Rha CK. 1978. *Gelatinisation of starch and wheat flour starch*. A review. Food Chemistry, 3 (4), 293–317.

This H. 2004. *Les gay-lussac, une sauce qui n'est pas classique*, https://pierregagnaire.com/pierre_gagnaire/travaux_detail/65, last access 2022-01-10.

As mentioned earlier, more or less once per month, I give to my friend Pierre Gagnaire, one of the most talented chef in the world, an "invention"; i.e., one application of a scientific idea. For gay-lussacs, in particular, it was based on the exploration of sauces.

This vo Kientza 2021. *Sauces*, in Handbook of molecular gastronomy, CRC Press, Boca Raton, 495–499.

Tovar J, Melito C, Herrara E, Rascon A, Pérez E. 2002. *Resistant starch formation does not parallel syneresis tendency indifferent starch gels*, Food Chemistry, 7(6), 455–459.

CHAPTER 11

More Complex Systems

11.1 HOW MUCH OIL IN A FRENCH FRY?

11.1.1 An Experiment to Understand the Question

11.1.1.1 *First, the General Idea*

It is said that French fries can be more or less easy to digest, depending on the quantity of oil that they "contain", but how much oil do they absorb generally? For studying this question, we could imagine to experimentally extract the oil from a fried French fry using an organic solvents such as ethyl acetate, and then evaporate the solvent to finally get the oil that would be weighted. This would work well, but the method would need a validation. A better way of doing this would be to conduct bibliographical research, in particular of validated methods, such as in AOAC documents (Srigley, 2017).

Here, we propose an easy experiment to explore this question (This, 1999; This vo Kientza, 2021a), based on the simple assumption that water boils when it is heated at more than 100°C, as it is the case in frying oil: if water from the outside of the potato sticks boils, streams of vapor are ejected. We can assume that oil cannot enter the fries during frying, so that if we eliminate the oil as soon as the potato sticks are out of the oil, they should have less oil inside.

Let us test this idea with the following protocol:

1. peel a potato;

2. cut it in rectangular solids about $10 \times 1 \times 1$ cm, trying to give them the same mass (use a scale to check this);

3. heat oil in a pan (be careful with very hot oil);

4. when the temperature is about 180°C, select two potato rectangular solids and record their masses;

5. prepare a large quantity of sheets of absorbing paper;

6. at the same time (exactly), put the two sticks in the hot oil and observe the phenomena: bubbles flowing out of the potato tissue, steam condensing on the

wall of a cold glass that you put over the oil bath, potato sticks first sinking, then floating, progressive disappearance, with bubbles flowing from places that goes on decreasing in numbers;

7. when the two French fries are golden in color, take them out of the oil exactly at the same time, sponge the first one carefully, while you simply deposit the other on a plate;

8. after 3 minutes, sponge the second French fry as carefully as for the first;

9. measure the mass of the two French fries;

10. repeat the experiments for other sticks, and calculate the standard deviation.

The last step is here for best practices and also for good training, because it invites one to make a statistical test. However, the result is clear even without such work: for a 10 g initial mass of potato, there is between 0.5 and 1 g of difference in mass, with a lower mass for the French fries that are carefully sponged immediately after frying. As the two French fries are submitted otherwise to the same process, it means that there is a difference in oil absorption; the oil remaining on the outside of the second stick being absorbed during the 3 minutes rest.

But let us interpret more thoroughly. At first, with the potato tissue (80% water) dipped in the oil at a temperature higher than 100°C, the water of the potato tissue boils, and this makes a large volume of steam (assuming that 18 g of water generates 30.5 L of steam, it can be calculated from the mass variation of French fries that about 1 L of steam is produced for one rectangular solid: this prevents oil from coming inside the fries).

The whole frying process does change visibly on the inside: if you cut a fried stick transversally, you will see a crust with both a purée and a space inside. After frying, when the fried potatoes sticks cool, the steam filling part of their inside condensates to liquid water. This sucks in the oil that was adhering to the outer surface (when the stick was not cleaned from its oil).

This experiment can be very important for "public health"... but also for our personal behavior. And we have to fear our own reflexes: a "triangle sensory test" (I don't want to explain it here, but the tests were blind) organized during one of the monthly seminars on molecular gastronomy in Paris showed that the jury was able to recognize the French fries which had been sponged from the more oily fries... but they preferred the fries with oil!

11.1.1.2 Questions of Methods

The general question of exchanges between the inside of food ingredients and their environment is very important for food science and technology, and one should know some general facts, in order to interpret them:

1. for meat (*i.e.*, animal tissues), collagen contraction after 55°C triggers the ejection of some liquid from the inside toward the environment: this is why the

mass of cooked meat is generally much less than the raw initial meat (a 1 kg piece can be reduced to 3/4 of its initial mass by roasting in an oven at 180°C);

2. when a food product is heated to more than 100°C, water is rapidly evaporated;

3. as a consequence, and contrary to what was sometimes taught in culinary schools, juices cannot go toward the inside of meat during roasting; on the contrary, they go outward. And this makes the brown residue when the water evaporates from these juices;

4. the same contraction of meat occurs during boiling, as can be easily seen by measuring the mass of meat; this is why there is no "expansion" of meat during boiling, *e.g.*, for making meat stocks;

5. plant tissues contain channels (xylem, phloem) for driving the raw sap up to the leaves, and the synthesized compounds to the lower part; various compounds can move toward these channels, or outside.

Now, a second methodological point: we have to insist that experiments have to be repeated, for "validation". In particular with food ingredients that can be both plant and animal tissues, the diversity between samples can be very large, and this leads to measurements with uncertainties. The general case can be depicted in Figure 11.1, and the comparison of values calls for statistics because there is a probability that the values are equal or not.

11.1.1.3 *Where the Beauty Lies*

Beauty can be physical or intellectual. And about potatoes, there are two wonderful ideas.

The first one is paradoxical: Alexander brings home 100 kg of potatoes (mathematically being made of 99% water). Overnight, the evaporation reduces the quantity of water until the proportion of water is 98%. What is the final mass? I am happy not to give the answer, but trust me: it is surprising.

Another funny idea about potatoes is related to the fact that professional cooks, when they make boiled potatoes, eliminate first the skin, cut the top and bottom, and cut the potatoes parallel to the long axis in order to make seven flat sides. Why seven? I invite you to consider the section that would be created with only three cuts: it would be triangular and a large proportion of material would be lost (have fun calculating it). If instead the cook makes four cuts, the section is a square, and the loss is less (by how much?). Calculate the loss as a function of the number of cuts, up to ten: Do you see something particular for seven?

Now, instead of calculating only the quantity of lost material, consider instead the quantity of loss avoided by one more cut: What do you see in the series of results?

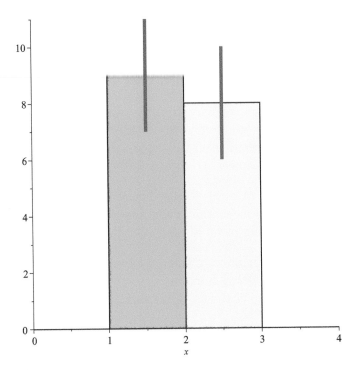

FIGURE 11.1 *Imagine that you weigh two objects three times each with a very uncertain balance, and that you get the average results 8 and 9. The large uncertainty (blue line) would prevent to consider that the masses of the two objects are different, and you can say that only in a probabilistic way.*

11.1.2 The Question

Let us come back to our frying question, and investigate the question of the quantity of oil that they can absorb. You remember that the real question is to invent a calculation about it. Many possibilities exist, but which one can we propose?

11.1.3 Analysis of the Question

11.1.3.1 The Data is Introduced

As for the experiment, we shall consider a stick of potato having a length of 10 cm, with a 1 cm × 1 cm section.

The proportion of water in potatoes is about 0.8 (Anses, 2022).

11.1.3.2 Qualitative Model

A potato is mainly made of water, plus starch, cellulose, pectins, and other organic compounds (FAO, 2008). When a potato stick is put into hot oil, water evaporates at the surface, forming bubbles that can be seen.

FIGURE 11.2 *A section of a French fries in the ideal case when the crust (brown) contains all the solid material of the potato, so that the inside is full of steam (blue); the surface of the fries is covered with adhering oil (yellow).*

In real French fries, the part at the outside makes a crust, the water being evaporated. During this part of the process heat is transferred to the inside, so that starch granules insides potato cells expand and gelatinize, making the equivalent of a purée. But more and more water escapes as steam, explaining why you see the purée and empty space when you transversally cut the French fries after frying.

Here, in order to make our calculation, we shall assume that all the solid content makes the crust, and that the inside is entirely full of steam (Figure 11.2).

When the fries are taken out of the oil bath, the steam re-condensates, and oil from the external surface is sucked in. Here, we shall calculate the worst case: fries cooling inside oil, so that all the steam within the crust would recondensate.

11.1.3.3 Quantitative Model

For this chapter let us go faster than usual, and let us do at the same time the quantitative description of the model, as well as the strategy for solving it.

In particular, each time a question is discussed, a preliminary choice has to be made, between the calculation of an order of magnitude, or a formal calculation.

Let us calculate first an order of magnitude.

We do that as follows:

1. The potato stick has an initial volume of about 10 cm^3 (1 by 1 by 10, in cm). For an order of magnitude, let us assume that it is entirely made of water.

2. We know that a small quantity of liquid water makes a large quantity of steam: at 100°C, about 20 g of water generate more than 30 L of steam, that is 1 g for 1 L (remember: only an order of magnitude). A consequence of this observation is that the volume of liquid water produced after recondensation of steam is negligible in the sticks: it will be full of oil, which makes more than 10 g of oil (we have seen that the density of oil is between 800 and 900 kg/m^3).

And that's it! By the way, calculation using orders of magnitude are very robust, so that indeed, they would be more useful for validating formal calculation than the reverse. But anyway, let us now do the formal calculation:

1. Assuming a constant size (no shrinking, no expansion during the frying process, because the rigid crust is made fast), we know the volume of the potato stick after frying.

2. The thickness of the final crust can be determined from the mass of dry matter in the potato: $1 - p$, if p is the water content for the fresh potatoes.

3. Of course, this mass has to be transformed in a volume, from the density of starch.

4. Knowing this, we can determine the volume v of the cavity full of steam: it is the initial volume minus the volume of crust.

5. And we can determine how much vacuum this steam will create by condensing: the volume of oil that we are looking for is equal to the initial volume minus the volume of crust and minus the volume of recondensed water.

11.1.4 Solving

11.1.4.1 Implementing the Strategy

1. We start from a stick of mass M. Remember that we want to calculate the volume of oil V_o given by the expression:

$$V_o = V - V_d - V_r,$$

where V is the initial (and final) volume of the stick, V_d is the volume of dried matter (mostly starch), and V_r is the volume of liquid water obtained by recondensation of the steam inside the volume $V - V_d$.

2. Because the initial mass is known (M) as well as the water content (p), we can determine the mass M_d of dry matter is:

$$M_d = (1 - p) \cdot M.$$

It does not change during frying.

3. As we know the density of starch ϱ_s, we can determine the volume corresponding to the mass of dry matter:

$$V_d = \frac{M_d}{\varrho_s} = \frac{(1 - p) \cdot M}{\varrho_s}$$

4. But we still do not know the initial volume of the stick. It is the sum of this dry matter volume plus the volume of water initially present in the potato material, *i.e.* the volume corresponding to a mass of water equal to $p\,M$:

$$V = V_d + \frac{p \cdot M}{\varrho_w}$$

where ϱ_w is the density of liquid water.

So that:

$$V_o = V - V_d - V_r = \left(V_d + \frac{p \cdot M}{\varrho_w}\right) - V_d - V_r = \frac{p \cdot M}{\varrho_w} - V_r$$

5. We now need to determine the volume of water that has recondensed from the volume initially occupied by water. To this end, we need to assume that the steam inside fries was at 100°C, so that we can use the information that 1 mole of liquid water transforms into 30.5 L of steam at 100°C.

We need now to know the number of moles of water in the volume $\frac{p \cdot M}{\varrho_w}$. Using the molar volume v of steam at 100°C ($v = 30.5\mathrm{e}^{-3}$ m³/mol), we can write:

$$n = \frac{\left(\frac{p \cdot M}{\varrho_w}\right)}{v}$$

As 1 mole of liquid water has a mass μ ($18\mathrm{e}^{-3}$ kg), the mass of liquid water that recondensate is:

$$M_r = n \cdot \mu = \frac{\left(\frac{p \cdot M}{\varrho_w}\right)}{v} \cdot \mu$$

And the corresponding volume would be:

$$V_r = \frac{M_r}{\varrho_w} = \frac{\left(\frac{\left(\frac{p \cdot M}{\varrho_w}\right)}{v} \cdot \mu\right)}{\varrho_w}$$

And now we have the volume of oil:

$$V_o = \frac{p \cdot M}{\varrho_w} - V_r = \frac{p \cdot M}{\varrho_w} - \frac{\left(\frac{\left(\frac{p \cdot M}{\varrho_w}\right)}{v} \cdot \mu\right)}{\varrho_w}$$

From which we find the mass of oil, using the density of oil ϱ_o:

$$M_o = \varrho_o \cdot \left(\frac{p \cdot M}{\varrho_w} - \frac{\left(\frac{\left(\frac{p \cdot M}{\varrho_w}\right)}{v} \cdot \mu\right)}{\varrho_w}\right)$$

11.1.5 Expressing the Results

11.1.5.1 The Formal Result Found is Written Down Again

Remember that we found:

$$M_o = \varrho_o \cdot \left(\frac{p \cdot M}{\varrho_w} - \frac{\left(\frac{\left(\frac{p \cdot M}{\varrho_w} \right)}{v} \cdot \mu \right)}{\varrho_w} \right)$$

In passing, let us observe that in spite of the temptation to use numerical values such as 30.5 (L), or 18 (g), we introduced symbols, in order to avoid absolutely mixing formal and numerical values.

11.1.5.2 Finding Digital Data

All values in the result were used before.

11.1.5.3 Introduction of Data in the Formal Solution

$$subs \left(p = 0.9, \varrho_w = 1000, \varrho_s = 1500, \varrho_o = 800, v = 30.5\mathrm{e}{-}3, \mu = 18\mathrm{e}{-}3, \right.$$

$$\left. M = 10\mathrm{e}{-}3, \varrho_o \cdot \left(\frac{p \cdot M}{\varrho_w} - \frac{\left(\frac{\left(\frac{p \cdot M}{\varrho_w} \right)}{v} \cdot \mu \right)}{\varrho_w} \right) \right)$$

$$7.20 \times 10^{-3}$$

This mass in kg corresponds to about 7 g. It is less than what we calculated first, but it is of the same order of magnitude. In order to interpret the difference, we have to remember that in the first calculation, we neglected the volume of dry matter, as well as the volume of water that recondensed from steam.

11.1.6 Conclusions and Perspectives

If you make the experimental test proposed in the introduction, you will find that only 1 g of oil is absorbed per stick, instead of 7 g, but this is already a lot, because (1) this is for only one stick of potato and (2) this is not safe oil, because it oxidized through heating! But remember the triangle test that was organized with French fries of the two kinds: oil absorbed or not (This, 2017). The fact that all members of the jury preferred the French fries with oil inside should probably be interpreted in terms of the biology of Evolution: human beings evolved so that finally they like fat because this biological material is important for living (in particular, it is useful to make the phospholipids of the membranes of our cells).

11.2 HOW THICK IS THE CRUST OF A SOUFFLÉ?

11.2.1 An Experiment to Understand the Question

11.2.1.1 First, the General Idea

Soufflés are culinary preparations with a reputation of being difficult to produce successfully, but this is because the cooks of the past did not know the correct rules that have to be applied. Using the following protocol, for a cheese soufflé, you will find it easy:

1. take a large bowl that can go in the oven, and coat the sides with butter;

2. now add a spoon of flour and shake the bowl so that the flour is distributed on all the sides as evenly as possible;

3. preheat the oven to 180°C;

4. in a pan, put a large spoon of butter, and heat it gently with two spoons of flour, until a light brown color is obtained;

5. add a glass of milk, salt, pepper and nutmeg to taste, and cook until the flour thickens the mixture;

6. let it cool; then add 100 g of grated cheese;

7. separately, take three eggs, and put the egg whites in another clean vessel, adding the yolks in the cold preparation obtained beforehand;

8. whip the egg whites until you get a firm foam;

9. add a small part of the foam to the content of the pan, and mix; then add all the remaining foam, and mix the whole preparation using a spatula (try to minimize the mixing process, so that you get a foamy preparation);

10. put the whole mass in the prepared bowl;

11. put the bowl on the heated bottom surface of the oven, and cook for about 30 minutes;

12. serve the expanded soufflé immediately: do not mind if it deflates: this is the hallmark of soufflés, contrary to cakes (Figure 11.3).

By the way, try to get a cheese with a lot of flavor, if you want to get a good soufflé: as for most preparations, you will not be able to cook good dishes with "bad" products.

11.2.1.2 Questions of Methods

Soufflés are prototypes of other culinary preparations that expand during cooking, such as choux, puff pastry, Durham popovers, some cakes, etc.

It is funny to remember that when molecular gastronomy was created in the 1980s, it was said in culinary schools that "eggs generate the expansion" in spite of

FIGURE 11.3 *For this soufflé, a crust was made under the grill before the ramekin was put in the oven. This make a flat upper surface, trapping partially the vapor bubbles.*

the fact that bread expands and does not contain eggs. It is also funny to remember that, at that time, it was written in important culinary books that soufflés and other preparations expand because of "the air dilatation by heating" (Montagné, 1934).

Is this really a good explanation, a good "theory"? Remember that the method, for sciences of nature, is to (1) observe a phenomenon, (2) characterize it quantitatively (you measure most aspects), (3) group the data in equations, (4) find a theory, with all equations (you have sometimes to introduce new concepts), (5) deduce refutable consequences of the theory, (6) make experimental tests in order to try to refute the theory (because a scientific theory is always insufficient, and the goal is certainly not to "demonstrate" it, even if we produced it ourselves, but to refute it, in order to find later a better theory).

Let us apply this method here. We know the phenomenon: it is the expansion of the soufflé, and the formation of a crust. Now, let us imagine the old culinary books did the job in order to produce the theory that they proposed. We have the burden to look for theoretical consequences and test them experimentally.

Let us observe that if thermal dilatation is concerned, we should be able to calculate it using the physical model about it, *i.e.*, the ideal gas equation:

1. before cooking, we apply the ideal gas model to the air inside the soufflé:

$$P_i \cdot V_i = nRT_i$$

2. after cooking, the same equation applies:

$$P_f \cdot V_f = nRT_f$$

3. let us divide the second equation by the first:

$$\frac{P_f \cdot V_f}{P_i \cdot V_i} = \frac{nRT_f}{nRT_i}$$

We are looking for the ratio of the final volume over the initial one:

$$\frac{V_f}{V_i} = \frac{P_i}{P_f} \cdot \frac{T_f}{T_i}$$

Now, we can assume that the pressure is constant because soufflés are quite soft, and the upper part would not resist pressure (indeed, the increase of the inner pressure is moving the upper part of the soufflé up, according this theory. This leads to:

$$\frac{V_f}{V_i} = \frac{T_f}{T_i}$$

Assuming a maximum temperature of 100°C (373 K) in the soufflé, we can calculate that the ratio is equal to 1.27: this mean an expansion by one third only, compared to 100–200% for real soufflés. Clearly, whereas it is true that air expands with heat, this expansion mechanism is not enough to explain the expansion of soufflés (This vo Kientza, 2021b).

11.2.1.3 Where the Beauty Lies

Finally, why do soufflés expand? If you look at soufflés during cooking through a glass window of the oven and a glass container for cooking the soufflés, you will be able to see bubbles moving upward and exploding at the surface. And if you weigh a soufflé before and after cooking, you will observe that the mass of the soufflé decreases during cooking: for example, a 100 g soufflé loses about 10 g. Obviously, this loss can only be water evaporating.

And this leads to an improved mechanism ("theory"), *i.e.*, that soufflés expand primarily because water evaporates: remember that the order of magnitude of vapor made from liquid water is more than 1 L for 1 g.

Of course, this has consequences: if the vaporization is from the top of soufflés, there is no expansion; on the contrary, if the soufflés are heated by below, then the steam bubbles form at the bottom of the cooking vessel, pushing the soufflé upward. Hence the rule: when you cook culinary preparations that have to expand, put them on the heated bottom of your oven.

11.2.2 The Question

Not all food cultures know what soufflés are, but when confronted with the question given in the title of this chapter, the students can go on the Internet in order to find that these dishes are traditionally made by cooking in an oven, at about 180°C, a mixture of whipped egg whites and a thick preparation, savoury or sweet, sometimes including flour (like cheese bechamel, thick pastry cream, or fruit puree). And the same question of expansion could be asked for bread, for cakes, etc.

Nonetheless, whether or not students know what a soufflé is, it is a fact that many find the question as difficult as the title of this chapter. Generally, they imagine that they will have to apply the Fourier equation about heat transfer, and they do not test more simple solutions. The solution looks like the one in the previous chapter; however, here we shall show that we should not be shy when confronted with such questions: let us choose the possibility that we prefer.

And now again, the question: how thick is the crust of a soufflé?

11.2.3 Analysis of the Question

11.2.3.1 The Data is Introduced

Only the recipe is given:

- 50 g butter

- 50 g flour

- cook until brown

- add 200 g milk

- cook until thickening of the "sauce"

- add grated cheese (to taste)

- cool

- add 4 egg yolks; mix

- salt, pepper qsp

- whip the 4 egg whites

- mix the whipped egg whites with the sauce

- put in a vessel

- cook 45 minutes at 180°C

11.2.3.2 Qualitative Model

The left part of Figure 11.4 shows the foamy preparation before being cooked. After mixing the thick sauce and the whipped egg whites, the air bubbles that are packed in the whipped egg whites are now distributed in the whole volume of the soufflé preparation; and this preparation is made of water and dry matter. During cooking, an expansion of the soufflé is observed because vapor bubbles form (and also air expands), and a crust is formed all around (right). This crust is the material of the soufflé deprived from its water.

FIGURE 11.4 *Formation of the crust of a soufflé. For the purpose of the calculation, one could imagine three steps: (1) the initial preparation, (2) its expansion, (3) making the crust from the outside layer. Of course, during real cooking the two processes occur simultaneously.*

11.2.3.3 Quantitative Model

Making a quantitative model begins by introducing symbols to describe the picture. As in all previous chapters, we have to consider both the physical and the chemical characteristics. Of course, during such a description, we risk introducing parameters that are useless, but this is not serious: if a parameter is useless, it will not be used; the effort of having introduced it is very limited.

Here, for sure, we are interested by the volume, the mass, the quantity of water.

Let V_1 be the initial volume of soufflé, and M_1 be its mass. This soufflé contains a mass $m_{w,1}$ of water, a mass $m_{d,1}$ of dry material, plus air.

During cooking, it expands, reaching a volume V_2.

11.2.4 Solving

11.2.4.1 Looking for a Solving Strategy

Many solutions can be proposed, but here we focus on one: we use the experimental result given above, of 10 g of water evaporated for a soufflé for which the initial mass is 100 g. The main idea of the calculation is to recognize that the crust is a part of the soufflé from which water is evaporated, as shown in Figure 11.4.

Let us make this slowly, because, as said, this exercise seems difficult for many.

1. We start from a preparation with water, dry matter and air.

2. A "density" parameter is introduced, as the mass of water per volume.

   (as seen in the middle picture of Figure 11.4).

3. In this expanded state, the mass of water does not change, and the "density" is changed (reduced, of course).

4. If you look at the middle part of Figure 11.4, you see that we divide the expanded soufflé in two parts:

   - the inner part,
   - the outer part, which will be reduced as the crust.

5. For the inner and the external parts, the water density is the same, before the evaporation of water.

6. We know that a mass of water evaporates. This value is used to determine the volume of the part that will make the crust.

7. We want now to determine how the external part is reduced through the evaporation of the mass μ of water (from middle to right part of the Figure 11.4).

   For this new calculation, we can use the same idea, but with a "density parameter" for dry matter.

8. And we can finish the calculation using the density (the real one) of dry matter ϱ, to determine the volume of dry matter in the crust.

9. This dry matter is distributed over the surface of the soufflé, so that we can determine the thickness.

11.2.4.2 Implementing the Strategy

As previously, we copy and paste the strategy, and we introduce the needed equations:

1. We start from a preparation with a volume V_1, a mass M_1, including water (mass m_w, as said above), dry matter, and air.

2. A "density" parameter is introduced, as the mass of water per volume:

$$\varrho_1 = \frac{m_w}{V_1}$$

3. This soufflé expands k times, the volume V_1 becoming kV_1 (as seen in the middle picture of Figure 11.4).

4. In this expanded state, the mass of water does not change, and the "density" becomes:

$$\varrho_2 = \frac{m_w}{kV_1}$$

(if you have trouble with that, imagine that the volume becomes twice the initial volume: now you understand more easily that the water density is divided by two).

5. If you look at the middle part of Figure 11.4, you see that we divide the expanded soufflé in two parts:

- the inner part, with volume V_i.
- the outer part, which will be reduced as the crust, loosing the mass μ of water, from a volume V_o.

6. For the inner and the external parts, the water density is the same, before evaporation of water. It is ϱ_2, as said above.

7. We know that a mass μ of water evaporates. This value is used to determine the volume of the part that will make the crust:

$$\varrho_2 = \frac{\mu}{V_c}$$

Thus:

$$V_c = \frac{\mu}{\varrho_2} = \frac{\mu}{\left(\frac{m_w}{kV_1}\right)}$$

And of course, the inner part has a volume:

$$V_i = kV_1 - V_c$$

8. We want now to determine how the external part is reduced through the evaporation of the mass μ of water (from middle to right part of the Figure 11.4).

For this new calculation, we can use the same idea, but with a "density parameter" for dry matter.

Before cooking, the mass of dry matter is m_d. The "mass of dry matter per volume" is :

mass of water per volume:

$$\varrho_{1,d} = \frac{m_d}{V_1}$$

After the expansion, it becomes:

$$\varrho_{2,d} = \frac{m_d}{kV_1}$$

This "density of dry matter" is the same in the inner part, in the external part, and in the whole soufflé, thus:

$$\varrho_{2,d} = \frac{m_d}{kV_1} = \frac{m_{d,e}}{V_c}$$

From which we find the mass of dry matter that will make the crust:

$$m_{d,e} = m_d \cdot \frac{V_c}{kV_1} = m_d \cdot \frac{\left(\frac{\mu}{\left(\frac{m_w}{kV_1}\right)}\right)}{kV_1}$$

9. And we can finish the calculation using the density of dry matter ϱ, to determine the volume of dry matter in the crust:

$$\varrho = \frac{m_{d,e}}{V_{d,c}}$$

Thus:

$$V_{d,c} = \frac{m_{d,e}}{\varrho} = \frac{\left(m_d \cdot \frac{\left(\frac{\mu}{\left(\frac{m_w}{kV_1} \right)} \right)}{kV_1} \right)}{\varrho}$$

10. This dry matter is distributed over the surface of the soufflé, so that we can determine the thickness.

Here, we need the surface of the soufflé: we assume that it is a cube, with a volume

$$V_i = kV_1 - V_c = kV_1 - \frac{\mu}{\left(\frac{m_w}{kV_1} \right)}$$

The side of this cube is:

$$c = \left(kV_1 - \frac{\mu}{\left(\frac{m_w}{kV_1} \right)} \right)^{\frac{1}{3}}$$

The area of the cube:

$$A = 6c^2 = 6 \left(kV_1 - \frac{\mu}{\left(\frac{m_w}{kV_1} \right)} \right)^{\frac{2}{3}}$$

From which we find the thickness of the crust using the equation:

$$V_{d,c} = A \cdot \varepsilon$$

That is:

$$\varepsilon = \frac{\left(\frac{\left(m_d \cdot \frac{\left(\frac{\mu}{\left(\frac{m_w}{kV_1} \right)} \right)}{kV_1} \right)}{\varrho} \right)}{6 \left(kV_1 - \frac{\mu}{\left(\frac{m_w}{kV_1} \right)} \right)^{\frac{2}{3}}}$$

11.2.5   Expressing the Results

11.2.5.1   The Formal Result Found is Written Down Again

$$\varepsilon = \frac{\left(\dfrac{\left(m_d \cdot \dfrac{\left(\dfrac{\mu}{\left(\frac{m_w}{kV_1}\right)}\right)}{kV_1}\right)}{\varrho}\right)}{6\left(kV_1 - \dfrac{\mu}{\left(\frac{m_w}{kV_1}\right)}\right)^{\frac{2}{3}}}$$

11.2.5.2   Finding Digital Data

Here, it is probably more interesting to discuss how one would find the needed data, than implement the final calculation.

For the determination of the water content, we need to analyze the recipe, ingredient by ingredient:

- 50 g butter: this means about 40 g of fat, 10 g of water;

- 50 g flour: no water;

- cook until brown: let us assume that no evaporation occurs;

- add 200 g milk: let us assume that it is about 200 g of water for simplicity;

- cook until thickening of the "sauce": same assumption;

- add grated cheese (to taste): let us admit that it does not change much the water content, because the quantity of cheese can be small;

- cool: no change in the water content;

- add 4 egg yolks (60 g water), mix;

- salt, pepper qsp;

- whip the 4 egg whites (110 g water);

- mix the whipped egg whites with the sauce;

- put in a vessel;

- cook 45 minutes at 180°C.

As a whole, it would make 380 g of water, out of a total mass of 540 g:

$$p = \frac{380}{540} = 0.8.$$

For the mass lost, if 100 g of soufflé preparation looses 10 g during cooking, so 380 g will loose 38 g of water.

Because 18 g (1 mol) makes 30.5 L (30.5×10^{-3} m^3), this will generate a volume of steam of

$$\frac{38}{18} \cdot 30.5 \, 10^{-3} = 6.44 \times 10^{-2} \text{ m}^3.$$

Of course, much of it is lost, and for the final volume V_f, we have to use experimental measurements, and assume it to be twice the initial volume. Let us observe that we do not know the initial volume, and we have now to calculate it, as the sum of the volumes of the ingredients plus the volume of the foam.

- if we use 50 g of butter, this makes about

$$\frac{50e{-}3}{800} = 6.25 \times 10^{-5} \text{ m}^3;$$

- for flour, we have also a volume

$$\frac{50e{-}3}{1500}; = 3.33 \times 10^{-5} \text{ m}^3;$$

- for milk, we have a volume of

$$\frac{0.2}{1000}; = 2.00 \times 10^{-4} \text{ m}^3;$$

- for the eggs, assuming eggs 60 g each and assuming that they are mainly made of water, this means a volume:

$$\frac{6 \cdot 60e{-}3}{1000} = 3.60 \times 10^{-4} \text{ m}^3;$$

- and for the foam, it can be seen in other chapters that we can assume a volume of 0.1e-3 m^3 for each egg white.

So that the total initial volume is:

$$\frac{50e{-}3}{800} + \frac{50e{-}3}{1500} + \frac{0.2}{1000} + \frac{6 \cdot 60e{-}3}{1000} + 4 \cdot 0.1e{-}3 = 1.06 \times 10^{-3}$$

Now, we can assume that the final volume will be double this one.

Finally, we do not know the density of the dry matter, and one has to consider that this is a mixture of fat, starch and proteins.

11.2.5.3 Introduction of Data in the Formal Solution

Would you like to do it?

11.2.6 Conclusions and Perspectives

The reason why we moved on rapidly is that I want to show that many solutions can be given to this question. Here is another one, that we find looking at the list of 14 definitions and equations: we see that it can be also a question of energy.

Let us use an oven of power P.

Its output (the fraction of the energy really given to the product) is rP, with $r < 1$.

The soufflé is cooked for a certain time s.

So that the energy given to the soufflé is:

$$E = Prt.$$

This energy is spent for:

- heating the soufflé, from 20 to 100°C

- evaporating water.

If M is the mass of the soufflé, and c_p its heat capacity (remember: we can find it as a weighted mean of the heat capacities of the ingredients), the energy needed for heating the soufflé from T_1 (remember: 20°C) to T_1 (remember: 100°C) is:

$$Q_1 = Mc_p\Delta T$$

Then the energy needed for evaporating a mass m of water is:

$$Q_2 = Lm$$

where L is the latent heat of evaporation of water.

Thus, finally:

$$rPt = Mc_p\Delta T + L \cdot m$$

From which we can calculate the mass m of water that evaporated.

11.3 HOW MUCH SALT IS THERE IN SPAGHETTIS?

11.3.1 An Experiment to Understand the Question

11.3.1.1 First, the General Idea

Often, it is advised to add salt to the water in which one cooks pastas, rather than salting the cooked pastas, but some cooks have the feeling that this amounts to spoiling the salt because a large part of it is lost in the water that is eliminated after cooking. First, is there a difference in the taste of pastas?

Here, we propose first to check that salt in water can be perceived:

1. in a pan, put 1 L of water and 10 g of salt (we follow an urban Italian rule: 1/10/100, which means 1 L of water, 10 g of salt and 100 g of pastas); bring to the boil;

2. at the same time, bring 1 L of water to the boil in another similar pan, but for this one do not add salt;

3. when water is boiling in both pans, add 100 g of dry pasta in each, and cook for 10 min;

4. when the time is over, immediately take the pastas out of the pan, and organize a triangle test:

   • use three identical plates, on the edge of which you mark numbers 1, 2, 3;
   • in two of them (chosen at random, 1 and 3, for example), put pastas cooked in salted water,
   • in the third plate, put pastas cooked with no salt;
   • propose people to test the three samples, with only one question: "Which samples are the same?"
   • now clean the plates, and put some salted pastas in plates 2 and 3, and non-salted pastas in plate 1, and make people test it as before.

This test will tell you if a difference is perceived.

When this first experiment is performed, compare pastas that are cooked either with salt in water, or with salt added after cooking. For a fair test, we have to add a carefully chosen quantity of salt, for the second batch (pastas cooked without salt, salt added afterwards).

Here, I propose that you use 2 g of salt, for the second batch (see more on this later). Of course, in order to have this amount of salt as evenly distributed as possible, I invite you to grind it (put it in a large pan, and rub it using a smaller one) in order to make what I called "icy salt" (the equivalent of icy sugar, but for salt).

And again, make a triangle test.

11.3.1.2 Questions of Methods

In the previous experiment, human beings are used as measuring tools. However, as any tool (think of a scale or a ruler), their precision is not perfect, in particular when the differences to measure are weak.

By the way, for such sensory tests, ask the jury to avoid looking at the pastas, because they could perhaps see the salt crystals, or a difference in color or brilliancy. For good methodology, a red light should be used, for example, and no communication between the members of the jury should be allowed. Be mindful also that the members of the jury give better results when focused (avoid disturbances).

Anyway when differences between samples are real but weak, there is a risk that the members of the jury may make a mistake. And also there is a risk that they can detect a non-existent difference: after all, a coin tossed many times sometimes falls on its head repetitively. One has to compare the result to a random result, using the following table:

Number of evaluations	Number of correct evaluations to reach a confidence of 95%	Number of correct evaluations to reach a confidence of 99%	Number of correct evaluations to reach a confidence of 99.9%
5	4	5	—
6	5	6	—
7	5	6	7
8	6	7	8
9	6	7	8
10	7	8	9
11	7	8	9
12	8	9	10

11.3.1.3 Where the Beauty Lies

Triangle tests are very useful, in gastronomical circles, because so many diverse —and often dubious— ideas have been around for a very long time: is it true that whipped eggs make a bigger foam when whipped always in the same direction? is it true that genoise are better when the batter is heated at 55°C? is it true that strawberries lose their flavor when washed, for example?

Indeed, testing these "culinary precisions" is an important aspect of the scientific discipline called molecular and physical gastronomy (for short "molecular gastronomy"). Now, with rigorous tests, a better culinary instruction can be given.

However molecular and physical gastronomy is not restricted to assessing phenomena: as in any science, its main goal is to look for the mechanisms of phenomena, and to this end, mathematical models have to be introduced. About spaghettis, one is interesting: fractals.

I do not want to explain this theory in detail, but only to show how it can tell us that spaghettis are cooked (This, 2009b).

The idea is to cook spaghettis, and take one at regular intervals (*e.g.*, any 20 s). Then, leave it from a certain height, randomly, on a white plate, and determine the "fractal dimension". For this, prepare a series of transparent slides on which you have drawn square grids of side 5 cm, 4 cm, 3, cm, 2 cm, and 1 cm. Put the grids one after other and count the number N of cells where the spaghetti is present. Then determine a quantity F equal to twice the logarithm of N divided by the square of the total number of cells. And look for the limit when the number of squares tends toward 0.

In this way, you will see that the limit increases with time, first rapidly, during the first 12 minutes, then slowly. Indeed, this reveals when the spaghettis are well cooked: between the two regimes.

11.3.2 The Question

How much salt is there in spaghettis?

11.3.3 Analysis of the Question

11.3.3.1 The Data is Introduced

No data is given.

11.3.3.2 Qualitative Model

Let us represent the making of spaghettis at the successive steps (Figure 11.5).
First, we put water in a pan.
Then, we add salt to water.
When salted water boils, spaghettis are added.
During cooking, the spaghettis swell, absorbing water and (possibly) salt.

11.3.3.3 Quantitative Model

We now do this formalization, picture after picture:

1. We see water, and this is a "material". The shape is not important, but we can describe it by:

   - its volume V,
   - its mass M,
   - its temperature T.

2. We see water plus salt.

   Let m_i be the mass of salt put in water.

3. We see now the spaghettis :

   We put n spaghettis in the pan.

   For a spaghetti :

   - length : L
   - radius : r_1

4. After cooking, let us assume that there was no evaporation (for example because there was a lid on the pan). This means that the mass of water remains M.

   Let the mass of salt remaining freely dissolved in water be m_f.

   Now, spaghetti swelled up:

   - same length L
   - radius r_2

FIGURE 11.5 *Here the colors are showing the different materials: changes of the color of water correspond to salt content changes. The spaghettis are represented by only one spaghetti (orange).*

The salt quantity in spaghettis is now μ

We do some hypothesis:

1. we do not consider the loss of starch in water;

2. we do not consider a possible saturation of salt, because this would mean 360 g per L, and this would be impossible to eat;

3. we assume that the water absorbed by spaghettis has the same concentration in salt than in the water outside the spaghettis (Sangpring et al., 2015.)

11.3.4 Solving

11.3.4.1 Looking for a Solving Strategy

We see that the quantity of salt in spaghetti is proportional to their volume increase, but only if the salt moves freely with water.

11.3.4.2 Implementing the Strategy

The mass of salt initially in water is m_i for a volume V of water. It means that in the volume equal to 1, the mass of salt would be $\frac{m_i}{V}$.

The spaghetti have an initial volume $n \cdot L \cdot \pi \cdot r_1^2$, and they have a final volume $n \cdot L \cdot \pi \cdot r_2^2$

This means that the volume of salted water that went into the spaghetti is:

$$n \cdot L \cdot \pi \cdot r_2^2 - n \cdot L \cdot \pi \cdot r_1^2$$

So that the mass of salt that migrated into the spaghetti is:

$$\mu = \frac{m_i}{V} \cdot \left(n \cdot L \cdot \pi \cdot r_2^2 - n \cdot L \cdot \pi \cdot r_1^2 \right)$$

11.3.5 Expressing the Results

11.3.5.1 The Formal Result Found is Written Down Again

For sure, we can put it exactly as we found it:

$$\mu = \frac{m_i}{V} \cdot \left(n \cdot L \cdot \pi \cdot r_2^2 - n \cdot L \cdot \pi \cdot r_1^2 \right)$$

But in view of estimating the result, let us factorize:

$$\mu = \frac{m_i}{V} \cdot n \cdot L \cdot \pi \cdot \left(r_2^2 - r_1^2 \right)$$

In this way, it is easier to recognize a volume in the last part of the expression (L multiplied by a square of radius). This makes μ homogenous as a mass, as it has to be.

We also see that the mass depends on the number of spaghetti pieces, and this is awaited.

Of course, μ is proportional to the initial mass of water, and this is fair because we assumed that the concentration of water that move into the spaghetti pieces was the same as the water in the pan.

11.3.5.2 Finding Digital Data

In SI units:

$$V = 1e{-}3 \text{ m}^3.$$
$$m_i = 10 \text{ g} = 10e{-}3 \text{ kg}$$
$$n = 50$$
$$L = 0.35 \text{ m}$$
$$r_1 = 1e{-}3 \text{ m}$$
$$r_2 = 2e{-}3 \text{ m}$$

11.3.5.3 Introduction of Data in the Formal Solution

$$subs\left(V = 1e{-}3, \, m_i = 10e{-}3, n = 50, r_1 = 1e{-}3, r_2 = 2e{-}3, L = 0.35,\right.$$

$$\left.\mu = \frac{m_i}{V} \cdot \left(n \cdot L \cdot \pi \cdot r_2^2 - n \cdot L \cdot \pi \cdot r_1^2\right)\right)$$
$$\mu = 1.65 \times 10^{-3}$$

This value is in the SI units. It is equal to 1.65 g, to be compared to the initial mass of salt added: 10 g.

11.3.5.4 Discussion, Playing with Parameters in Order to Explore the Solution Space

For validating this result, let us calculate an order of magnitude. We assume that pastas double in diameter during cooking: their volume is multiplied by about four. With a mass of 100 g of dry pastas and a density of 1500 kg/m^3, this makes an initial volume of $\frac{0.1}{1500} = 6.7e{-}5$ m^3.

If the volume was increased fourfold, this means that the volume of absorbed water is three times the initial volume, $i.e.$, 2.01e-4 m^3. Now, let us assume that the salt of the pan is absorbed with the same proportion as water, as we used 10 g for 1e-3 m^3, this means that the quantity of salt to add is 2.01 g.

And now, let us play with the result, to explore various cases:

1. For example, let us double the quantity of salt added :

$$subs\left(V = 1e - 3, \, m_i = 20.10^{-3}, n = 50, r_1 = 1e - 3, r_2 = 2e - 3, L = 0.35,\right.$$

$$\left.\mu = \frac{m_i}{V} \cdot \left(n \cdot L \cdot \pi \cdot r_2^2 - n \cdot L \cdot \pi \cdot r_1^2\right)\right)$$
$$\mu = 2.03 \times 10^{-5}$$

Of course, the quantity of salt in the spaghetti is doubled.

2. Let us now imagine that the quantity of water is reduced:

$$subs\left(V = 0.5e{-}3, m_i = 10e{-}3, n = 50, r_1 = 1e-3, r_2 = 2e-3, L = 0.35,\right.$$

$$\left.\mu = \frac{m_i}{V} \cdot \left(n \cdot L \cdot \pi \cdot r_2^2 - n \cdot L \cdot \pi \cdot r_1^2\right)\right)$$

$$\mu = 3.30 \times 10^{-3}$$

Here the increase of the salt content is increased as well.

3. What about changing the swelling of spaghetti, *e.g.*, making the final radius three times the initial one?

$$subs\left(V = 1e-3, m_i = 10e-3, n = 50, r_1 = 1e-3, r_2 = 3e{-}3, L = 0.35,\right.$$

$$\left.\mu = \frac{m_i}{V} \cdot \left(n \cdot L \cdot \pi \cdot r_2^2 - n \cdot L \cdot \pi \cdot r_1^2\right)\right)$$

$$\mu = 4.40 \times 10^{-3}$$

Now, another increase is observed.

11.3.6 Conclusions and Perspectives

The quantity of salt is small in spaghetti (fortunately, because salt can be harmful in excessive quantities). It can be observed that this calculation assumes that the salt migrates into spaghetti as water does. This should be certainly considered more thoroughly, first by a thorough research of the bibliography (rice can be considered as well, because it includes starch granules, as for spaghettis), but also a practical session can be organized for the experimental determination.

REFERENCES

Anses. 2022. *New potato, raw*, Ciqual, https://ciqual.anses.fr/#/aliments/4023/new-potato-raw, last access 2022-01-26.

Here, the database is giving a water content of 79.9 g/100 g... but beware: plant and animal tissues are always very diverse.

Burke R, Kelly A, Lavelle C, This vo Kientza H (eds). 2021b. Handbook of Molecular Gastronomy, CRC Press, Boca Raton, USA.

FAO. 2008. *Potatoes, nutrition and diet*, http://www.fao.org/potato-2008/en/potato/factsheets.html.

Montagné P. 1934. *Larousse gastronomique*, Larousse, Paris.

Sangpring Y, Fukuoka M, Ratanasumawong S. 2015. *The effect of sodium chloride on microstructure, water migration and texture of rice noodle*, LWT-Food Science and Technology, 64, 1107–1113.

Srigley CT, Mossoba MM. 2017. *Current analytical techniques for food lipids*, Food and Drug Administration Papers 7, https://digitalcommons.unl.edu/cgi/viewcontent.cgi?articgle=1011&context=usfda.

This H. 1999. *Experiment for the Science TV Programme of NHK*. Japan.

This H. 2009a. *Fractales et spaghettis*, Pour la Science, 384, 99.

This H. 2017. *Compte rendu du séminaire de gastronomie moléculaire*, October 2017. http://www2.agroparistech.fr/IMG/pdf/sem_oct_17_.pdf, last access 2018-12-15.

Each month (except in July and August) since September 2000, we have a "seminar on molecular gastronomy", during which we make experimental tests of culinary precisions (tips, methods, old wive tales, etc.). You can attend and apply for free to receive the reports at icmg@agroparistech.fr.

This vo Kientza H. 2021a. *How to reduce oil in French fries: a student experiment*, Handbook of molecular gastronomy, CRC Press, Boca Raton, USA, 663–664.

CHAPTER 12

Conclusion

T HE POSSIBILITIES OF calculation in the kitchen are infinite, from the simplest (*i.e.*, extending a recipe to a different number of guests, simply by applying proportionality) to the more complex (remember the discussion about the application of the Maxwell-Boltzmann distribution to the sedimentation of sucrose in water). Anyone can play the game of first conducting an experiment, to test ideas, and then creating a model and performing a calculation, transforming one's knowledge into skill.

This bears repeating: anyone can tackle the preceding questions, and others, by using one's knowledge. If you know only the proportionality, play this game of calculating proportions. So far, so good. But if you know quantum mechanics, why couldn't you try to use it also? Indeed, this is exactly why sciences of nature are so wonderful: experiments lead to calculations, in order to see behind the surface, to see deeper and deeper.

Certainly, some people who do not excel at calculation sometimes criticize the theoretical basis behind the evidence; and some others criticize the simplified models, accusing them of distorting the reality, of being "simplistic"... this need not concern us! Sciences goes to the root of phenomena, using the wonderful intellectual tools that were introduced by our predecessors: energy! momentum! force! inertia! chemical potential! entropy! atoms! molecules!

Certainly, rationality needs training, and this training is based on the sound consideration of numbers and equations. In science, our words do not lie: they have only one meaning, and this meaning is quantitative. Our theories are not fantasies, but quantitatively compatible with measurements. More and more precise measurements. Yes, we live sometimes in another world, but reminiscent of the Middle Age, where a correspondence was assumed between our world and the heavens, we are on two levels: abstraction and experiment.

Here, we limited ourselves to what happens in the kitchen, but this is no small world. On the contrary. Cooking being a "chemical art", we find remarkable application of chemistry, on the one hand and physical chemistry on the other. And this leads me to apologize for my own passion: to be sure, in this particular book, the chemistry was but a small part. This is for next time.

DOI: 10.1201/9781003298151-12

Index

Note: Locators in *italics* represent figures and **bold** indicate tables in the text.